Surface Active Agents

The production and use of surface active agents have seen various evolutions over time, yet rarely, if ever, has this information been collated in one place.

Covering all surfactant classes in a clear and concise style, from their properties and applications to an overview of the evolution of their production processes, this book is a comprehensive overview of the field. It is both a record of important documents and intellectual property as well as a springboard for possible future developments.

Key features:

- Covers both man-made and natural surfactants
- Includes abundant references to production processes and developments of intellectual property
- Provides a complete background to the field of surface active agents today

From producers and formulators of surface active agents to professors and students of raw materials, this book is appropriate for both academic courses and industry professionals.

Dr. Guido Bognolo has a degree in theoretical chemistry and has spent his professional career working on the technology, applications, marketing, strategy, and business acquisition of surface active agents in multinational chemical companies. He was for many years the managing director of WSA Associates, a company providing consulting services for technology and strategic investments to several companies operating in the field of surface active agents and specialty chemicals. As general secretary of the Confederation of Senior Expert Services (CESES) he executed many consulting missions in the People's Republic of China. He holds a master's degree in business administration from the Solvay Business School of Economics and Management in Brussels. He is the author of several publications on the technology, marketing, and strategy aspects of surface active agents, as well as chapters in reference books.

Surface Active Agents

Historical Perspectives and Future Developments

Guido Bognolo

CRC Press
Taylor & Francis Group
Boca Raton London New York

CRC Press is an imprint of the
Taylor & Francis Group, an **informa** business

Cover image credit: © shutterstock

First edition published 2024
by CRC Press
6000 Broken Sound Parkway NW, Suite 300, Boca Raton, FL 33487–2742

and by CRC Press
4 Park Square, Milton Park, Abingdon, Oxon, OX14 4RN

CRC Press is an imprint of Taylor & Francis Group, LLC

© 2024 Guido Bognolo

ISBN: 9781032517681 (hbk)
ISBN: 9781032517735 (pbk)
ISBN: 9781003403869 (ebk)

DOI: 10.1201/9781003403869

Typeset in Times
by Apex CoVantage, LLC

Historia magistra vitae

(History is a wonderful teacher/History shows the way)
Ancient Roman saying

Contents

Foreword xi
Preface xv
Acknowledgements xvii
Introduction xix

PART I   Natural Surface Active Agents

Chapter 1   Soap, the First Man-Made Surface Active Agent 3

Chapter 2   Soap-Making Process and Raw Materials 19

              2.1   Oils, Fats, and Fatty Acids 19
              2.2   Alkali 19

Chapter 3   Later Soap Developments 23

Chapter 4   Naturally Occurring Surface Active Agents 25

              4.1   Introduction 25
              4.2   Saponins 25
               4.3   Bile Acids 27
              4.4   Phospholipids 27

PART II   The Advent of Synthetic Surface Active Agents

              Introduction 29

Chapter 5   Amphoteric Surfactants 31

              5.1   Introduction 31
               5.2   Amphoteric Surfactants Production Process and
                     Raw Materials 32

Chapter 6   Anionic Surfactants 35

        6.1   Introduction 35
        6.2   Sulphated Oils and Fatty Acids 35
        6.3   Sulphated Alkyl Esters 36
        6.4   Sulphated Glycerol Esters 36
        6.5   Sulphated Amides 37
        6.6   Sulphated Olefins 38
        6.7   Sulphated Fatty Alcohols (Alkyl Sulphates) 38
        6.8   Sulphated Alkyl, Aryl, and Alkylaryl Ethers 38
        6.9   Fatty Alcohols Sulphation Process and Raw Materials 39
        6.10 Alkylaryl Sulphonates 40
                6.10.1 Alkyl Naphthalene Sulphonates 40
                6.10.2 Naphthalene Sulphonate-Formaldehyde
                        Condensates 41
                6.10.3 Alkyl Benzene Sulphonates 41
                6.10.4 Alkyl Diphenyl Oxide (Di)Sulphonates 43
        6.11 Alkyl (Paraffin) Sulphonates 44
        6.12 Olefin Sulphonates 45
        6.13 Methyl Esters Sulphonates 46
        6.14 Isethionates 48
        6.15 Dicarboxylic (Sulphosuccinates) and Tricarboxylic
               Sulphonated Esters 49
        6.16 Sulphonation Process and Raw Materials 50
                6.16.1 Alkylation of an Aromatic Moiety 50
                6.16.2 Sulphation and Sulphonating Agents 51
        6.17 Phosphate Esters 53
        6.18 Surfactants Based on Amino Acids and Protein
               Hydrolysates 56
                6.18.1 Acyl Taurates 57
                6.18.2 Acyl Sarcosinates 58
                6.18.3 Acyl Glutamates 58
                6.18.4 Acyl Glycinates 59
                6.18.5 Other Amino Acid Surfactants 59
                6.18.6 Hydrolysed Proteins Surfactants 60

Chapter 7   Cationic Surfactants 67

        7.1   Introduction 67
        7.2   Cationic Surfactants Production Process and
               Raw Materials 67
                7.2.1 Quaternary Ammonium Compounds 67
                7.2.2 Ester quats 69

Chapter 8   Nonionic Surfactants 71

        8.1   Introduction 71

8.2 Alkoxylation of Fatty Alcohols and
        Alkyl Phenols.. 71
8.3 Fatty Acid Esters of Polyhydric Alcohols and Their
        Alkoxylated Derivatives... 72
        8.3.1 The HLB Concept.. 73
8.4 Sorbitan Esters 74
8.5 Polysorbates 75
8.6 Alkylpolyglucosides 75
8.7 Amine Oxides... 76
8.8 Ethoxylated Fatty Amines 78
8.9 Ethoxylated Methyl Esters 78
8.10 EO, PO, and EO/PO Homo- and Copolymers.......................... 79
8.11 Ethoxylated Mercaptans.. 80
8.12 Nonionic Surfactants Production Process and Raw
        Materials 81
        8.12.1 Alkylphenols and Alkyl Naphthols............................ 83
        8.12.2 Fatty Alcohols... 83
                8.12.2.1 Natural Fatty Alcohols 84
                8.12.2.2 Synthetic Fatty Alcohols 85
        8.12.3 EO 89
        8.12.4 PO 91
        8.12.5 Fatty Acids 91
        8.12.6 Fatty Amines... 91

Chapter 9 Other Surfactants 97

9.1 Silicone-Based Surfactants 97
9.2 Fluorosurfactants 98
9.3 Acetylenic Surfactants 99

PART III   Washing and Cleaning Habits

Chapter 10 Washing Linen and Clothes 103

Chapter 11 Body Washing and Personal Hygiene 105

Index.. 115

Foreword

Dr. Guido Bognolo's book can be recommended not only to those interested in the history of science, but it is relevant also to those of us who want to learn more about how some of our habits have evolved. The development of surfactants is intimately linked to how fundamental human activities such as washing clothes, body cleaning, and painting have evolved. This is a fascinating angle of history.

It is likely that human beings always have wanted to clean their garments but until quite recently the methods used were extremely primitive. From ancient times until the beginning of the 20th century, the main tool was alkali, often potash, but the Romans also collected and used urine, which is not alkaline, for the purpose, as one can read about in the book. Soap was invented in the Middle East, maybe in ancient Babylon. It was probably produced by boiling animal fat together with a slurry of wood ash. Ash is alkaline, so this procedure is what today would be referred to as saponification, that is, alkaline hydrolysis of a triglyceride to yield fatty acid soaps together with glycerol. The fatty acid soap is a surfactant, still the main constituent in common toilet soaps. Soap bars are in principle made by the same procedure today as the Babylonians used some 5000 years ago, although animal fat has largely been replaced by vegetable oils and the alkali source is now a more well-defined product than wood ash.

Soap can be regarded as the first surfactant and has been known for thousands of years. However, it was not extensively used for garment washing. The washing of garments through history was more a mechanical than a chemical procedure, as one can read about in Dr. Bognolo's book. Dirt was removed from clothes and linen by smashing the wet garment on rocks or beating it with sticks, indeed a cumbersome activity. Things changed completely first when running water became available in the cities, then when modern detergents were introduced – Henkel's Persil, launched in 1907, was the first – and then again when household washing machines became available at a reasonable price.

It is also likely that human beings always have had a desire to wash themselves, although certainly not to the extent we do today. The habit of bathing, in ancient Rome and in many other civilisations, is well documented, and there is a long chapter in the book that describes bathing habits in different cultures. The issue of nudity, as well as of whether there should be separate baths for men and women, was a difficult matter, handled differently in different places and at different times. The latter issue is a problem that we see again in Europe now with the growing Muslim population. The bathing as such is well documented but much less is known about whether those who bathed, or cleaned themselves in other ways, used "chemicals" to remove the dirt. As mentioned earlier, soap was known for a long time. However, it is not clear to what extent it was used for general body cleaning in ancient times and during the Middle Ages. What is well documented is that people used fragrances to mask the odour. Knowledge of perfumery came to Europe in the 14th century, but perfumes have a much longer history than that. The Romans, the Persians and the Arabs all used perfumes 2000 years ago. There is a long chapter in the book related to the history of body washing and personal hygiene.

Another habit in which the use of surfactants dates back a long time is painting. Humans have always had a desire to paint and decorate. We can still witness this in caves around the world. and historians believe that fabrics of different types were often colourfully decorated. It is likely that body painting also existed far back in history, predating today's tattooing habit by many thousands of years. Without surface active agents of some kind, painting would have been difficult. Surface active agents are needed to disperse the pigments and sometimes also to wet the surface to be painted. Dyeing woollen clothes constituted a special problem. The fresh wool is covered by a thick layer of fat which must be removed prior to dyeing. For this, soap was used.

The book covers essentially all commercial types of surfactants. It presents the history and the key patents involved. It also briefly discusses the characteristic properties and main uses of the different surfactants. It conveys practically useful information, such as production aspects and product costs. It touches on environmental issues, but the rate of biodegradation and aquatic toxicity, which are today very important parameters in the choice of surfactant for various applications, are not dealt with in detail.

Not only does the book cover different aspects of surfactants, but it also includes the most important surfactant raw materials, such as natural and synthetic fatty alcohols, alkylphenols and alkyl naphthols, fatty acids, and ethylene and propylene oxide, among others. This is very useful because this type of information is not easily available in a comprehensive form elsewhere. As an example, the chapter on synthetic fatty alcohols contains the history behind and the chemistry involved for (1) the Shell higher olefin process, (2) Sasol's Fischer–Tropsch olefin process and (3) the ALFOL and the EPAL processes, sometimes collectively referred to as Ziegler processes. The different processes give different branching patterns, and everyone involved in formulation work based on surfactants knows that the structure of the hydrophobic tail of the surfactant, not only the number of carbon atoms, is important for the self-assembly in bulk and adsorption at surfaces.

Overall, the book gives a good account of the innovation ability and the massive chemical engineering that lie behind the development of today's surfactants. Europe, and particularly Germany, had a lead in this development, but a substantial part of the inventions occurred also in the US. One country outside Europe and the US that stands out as a major contributor to this development is South Africa. During the time of the apartheid, when the country's supply of oil and gas was blocked, the country made a massive effort to develop the German Fischer–Tropsch process from the 1920s to make liquid hydrocarbons from hydrogen and carbon monoxide. The gaseous feedstock was obtained by the gasification of coal. Synthetic fatty alcohols could then be obtained by hydroformylation of olefins followed by hydrogenation. The company later extended the technology to make highly branched alcohols via the Guerbet reaction and to nonionic surfactants; that is, they reacted the alcohols with ethylene oxide. Sasol has subsequently been a successful company, and after a number of acquisitions, it is now the largest producer of nonionic surfactants in the world. Its headquarters are still in South Africa.

Dr. Bognolo is not entirely enthusiastic about the development of surfactants. On the contrary, he is deeply concerned about the lack of innovations during the last

50 years. The golden age of surfactants was the period from 1920 to 1970. (To a large extent, that was the golden age of chemistry as a whole.) Most of the important types of surfactants that we use today appeared on the market during this half century. The important processes involved in surfactant production, such as the alkylation of aromatics, the development of processes to make synthetic alcohols, sulphonation, sulphation, alkoxylation, and so on, were developed and commercialised during this period. After 1970, only a few new products have reached the market at a large scale, with the paraffin sulphonates and the ester quats being the most important exceptions.

There have been substantial innovations also during the last half century related to silicone surfactants and fluorinated surfactants, but these are small-volume specialty surfactants. In addition, they will probably experience increasing problems in the future because they face problems in meeting increasingly strict environmental regulations.

The environmental regulations, which focus on rate of biodegradation and aquatic toxicity, is today a considerable obstacle for the development of new surfactants. The regulatory costs for bringing a new surface active molecule to the market is very high, and it is understandable that companies think twice before they enter such projects. Chemists in general sympathise with the environmental regulations set up by the Organisation for Economic Co-operation and Development, but one must then be prepared to pay the price of an almost stand-still when it comes developing radically new surfactant molecules.

Thus, the picture that Dr. Bognolo conveys about surfactants is one of a sunset industry. All major developments have been done, the products are old and the market is not increasing much. Dr. Bognolo writes: "I am still asking to myself why an explosive and innovative expansion faded away in an almost inexistent innovative horizon". Personally, I agree with this view. The surfactants that we use today were developed many decades ago, and the pipeline is rather empty. However, surfactants can be combined in different ways, and it is the combination of surface active molecules that is the product that the customer sees. Combining surfactants is part of what is called formulation science. During the last half century, when we have seen little in terms of completely new surfactants, we have instead witnessed major steps forward in formulations, involving combinations of surfactants and/or combinations of surfactants and amphiphilic polymers. The surfactant world has changed from being a playground for organic chemists to being an arena dominated by physical chemists.

Gothenburg, Sweden, November 2022

Professor Emeritus Krister Holmberg
Chalmers University of Technology
Chemistry and Chemical Engineering
SE-412 96 Göteborg
Sweden

Preface

The idea of writing an essay on the history of surface agents originated from two considerations.

The first was the dismaying absence in recent years of true innovation in surface active agents. Over two, if not three, generations not much substantially novel has emerged. That, whether you like it or not, it is a reality that the surface active agents of general consumption today were invented in the 1930s, 1940s, and a little bit in the 1950s. Admittedly in those years, researchers were not constrained by burdening regulations and mass consumer influences, so they had plenty of space to express their creativity, albeit this often took the form of a "meccano chemistry" approach. By this I mean the habit of synthesising surface-active molecules by linking hydrophobes and hydrophiles in a matrix-structured pattern. It produced a huge number of molecules that were looking for applications, rather than addressing the shortcomings of the products available at the time, and use sound, scientific principles and rules of colloid science to design "ad hoc" problem-solving structures.

But I am still asking myself why an explosive and innovative expansion faded away in an almost inexistent innovative horizon. Maybe it is because meccano chemistry produced so many structures, of which only a handful reached commercial success. It is like a mindset of "all that could be synthesised has been synthesised" began to pervade the academia and the industry. Maybe it is the ever-increasing regulatory burden that resulted in unbearable registration costs; maybe it is public opinion and opinion-forming influencers that made certain raw materials and surface active products unpalatable to a large number of consumers. More likely it is the combination of these and possibly other influences, but we have to accept that the surface active realm has slipped into a dismaying stagnation. In this situation, maybe taking a look at the past, not only in terms of realisations but also of mindsets, may awaken a sleeping beauty.

Second, there is a wealth of knowledge that was produced in the golden years of the surface active agents and that risks falling into oblivion (and in many instances, it has already met this fate). It is problematic to retrieve the original works in today's overflow of information (by the way, many of dubious quality) that favours just the recent present. I thought that this work may help save at least some of this very valuable heritage.

What is presented is limited by the resources at my disposal, from the bibliographic research to the writing and editing of the text. I have done my best to be accurate and precise and I apologise up front for any inaccuracies. Any observations are welcome.

Meanwhile, I hope the reading of these notes will be instructive and hopefully pleasant.

Acknowledgements

I am deeply indebted to Dr. Joseph C. Dederen and Dr. Derrick C. Dobson for the reviews, suggestions and advice that enabled the completion of this work; Professor Krister Holmberg for his preface; and Dr. Ing. Giulio Groppi for organising the text in a form as demanded by the editor.

Introduction

Despite the amazing diversity of structures and effects, surface active agents have a common denominator that makes them different from any other chemical: each molecule mimics, in practice, two molecules, one having a pronounced affinity for low-polarity substances and the other being comfortable only in high-polarity media. It is the combination of these two moieties in one single structure that enables surface active agents to make incompatible phases coexist in homogeneous systems, often of considerable complexity and, if properly formulated, isotropic and thermodynamically stable.

Surface active molecules may have been used, albeit unconsciously, since humankind evolved the first social structures and began to develop cultures, allowing for more elaborate activities than just raising food, for example, painting on rocks, making pottery, and body beauty.

The first known evidence for the deliberate planning, production, and curation of a pigmented (red ochre) compound was found in the Blombos cave in the Blombos Private Natural Reserve about 300 kilometres east of Cape Town and dated to between 100,000 and 70,000 years ago. Although no fats, resins or waxes have been positively identified (these may have degraded in the course of the millennia), it is conceivable that such materials were used in the preparation of the compounds. It is possible that pyrotechnology was used to facilitate the extraction of fats from animal offal. It is to be expected that some primordial recipes were followed and transmitted (1). It is obviously impossible to know for what purpose the pigmented compounds were produced, but an accredited hypothesis is that they were probably used for ritual or symbolic purposes.

Chemicals with surface activity are abundantly present in nature. Besides oils and fats that can be converted into fatty acids and soaps, there are saponins, phospholipids, rosin esters and fatty acids, egg yolk and other proteins and polysaccharides. It is reasonable to suppose that the accidental use of vegetal and animal products containing such substances enabled the observations of their effects and sparked the conscious use for specific purposes.

Colouration and decoration are very sense-appealing and something that must have surely caught the attention of our prehistoric ancestors, stimulating their imagination and creativity. Any form of painting, be it on the walls of the Altamura or Lascaux caves or for body beauty or for pottery decoration, needed pigment dispersion and binding, and these effects were achieved with substances available from natural sources. Still early in the 20th century, tribes that have retained habits reminiscent of prehistoric times (e.g., in southern Africa) used animal blood to bind pigments in their wall painting preparations. It is likely that egg yolk or other polymeric natural binders were a formulation ingredient of the inks Egyptian scribes used already in pre-dynastic times. Here we have already a highly sophisticated civilisation that had expressed outstanding architectural achievements and developed more down-to-earth technologies to make cosmetics, colour linen and clothes, brew beer and cook food, many demanding surface active substances and effects.

It is commonly accepted that soap, the surface active agent "par excellence", was first discovered probably 5000 years ago, most likely in Mesopotamia. An explanation put forward is that the fat dripping from roasted animals was saponified on contact with the alkaline ashes onto which it dropped. The amorphous, wax-like substance was probably found to have interesting effects, which may have prompted its recovery and storage, as well as the development of specific production procedures.

After the early start of the soap industry and the use of other natural surface active agents, there were no significant developments worth reporting for several hundreds of years, and one may rightly wonder why, considering the evolution in practically all domains of human activity. The answer is that until the beginning of the last century, the driving forces for innovation and change were not at work at the same time. In the case of surface active agents, these can be summed up as follows:

- The availability of raw materials
- The demand for new or enhanced effects
- The availability of technologies to deliver such effects
- The existence of economic conditions enabling the payment for the effects

First, there was no surplus of fats as the productivity of farming and breeding was low and barely sufficient to feed a growing population. Second, despite the demographic expansion, not many people were interested in personal or garment cleaning or were prepared to pay for it. Do not forget that well into the 18th century, people did not consider cleanliness a positive virtue and bathing a social norm. Finally, there was no real demand for effects other than the ones provided by soap, and even assuming there were some, the chemical and manufacturing technology was inadequate for their delivery.

Then in the second decade of the last century, all the conditions for accelerated technological and commercial development converged to fuel an exponential expansion:

- The search for alternative raw materials, originally driven by strategic and military considerations met with the availability of suitable substrates and with the synthetic and manufacturing technologies enabling to turn them into functional products at good profitability.
- The new manufacturing processes from a rapidly expanding industry demanded enhanced interfacial effects.
- A combination of socio-economic factors led both to higher hygiene standards and to the possibility to pay for them.

A major driving force was the availability of raw materials, in the context of utilisation of waste by-products and of extraction of higher added value from commodities. This was at the basis of developments like the sorbitan esters and the polysorbates (sorbitol and ethylene oxide), paraffin and olefin sulphonates (n-paraffins and alpha-olefins), alkylbenzene sulphonates (propylene, benzene) alkyl phenols alkoxylates (propylene, phenol, ethylene and propylene oxide), alkyl polyglucosides (glucose, molasses, starch), sugar esters and sucroglycerides (sucrose, fats, molasses), synthetic fatty alcohol alkoxylates (α-olefines, ethylene, ethylene and propylene oxide), glycerol esters (glycerol, fats), ethylene oxide and propylene oxide adducts (ethylene and propylene).

For many years, the growth of the surface active market outpaced the gross domestic product (GDP) expansion, driven by

- The new industrial processes emerging as a consequence of the industrial expansion and of the research efforts during and in between the two World Wars.
- The recognition that surface active agents could greatly improve the process yields, the product quality, the economics of manufacturing or all of those at once.
- The evolution of washing habits (fabric and personal) and the great simplification of domestic laundry brought about by washing machines and products.

This reached its climax in the late 1950s/early 1960s when the research of substances with amphipathic properties produced an amazing number of synthetic products. We witnessed then the appearance of sulphoxides; phosphine oxides; benzyl-, allyl- and chloride-capped ethoxylates; sulphonamides; carbamates; guanidines; phosphonamidic acids; ethoxylated mercaptans; sulphobetaines; borate esters; guanyl and acylguanyl ureas; alkoxylated dehydroabietylamines; and polyoxyethylene t-aliphatic alkylamines, just to quote a few. Eventually, natural selection exacted its toll, and practically none of these developments survived the scrutiny of cost-effectiveness, outperformed as they were by products simpler in conception and easy to manufacture from cheap raw materials with comparatively inexpensive processes, for example, alkylbenzene sulphonates, alkylphenols ethoxylates, fatty alcohols alkoxylates, and sulphates introduced commercially already in the 1930s and 1940s.

A confirmation is that reference books of the 1950s and 1960s barely mention amphoteric and cationic surfactants. The lion's share in those years was taken by the alkyl aryl sulphonates and far behind by the nonionics represented by alkyl phenol and fatty alcohols ethoxylates.

In the second half of last century, there have been comparatively few original developments of surface active molecules: commercially, there was, in the 1970s, the resurrection and broad introduction of the paraffin sulphonates (by the way, nearly three generations after the first laboratory synthesis and more than one generation after their initial commercial exploitation) and, in the 1980s, the introduction of the quaternary ammonium esters in response to the concern on the environmental impact of the alkyl ammonium derivatives hitherto used in fabric softening formulations. Note that quaternary ammonium esters had been already synthesised in the 1930s, however without generating a commercial interest.

In this period, true technological innovations came from the silicone surfactants, the acetylenic surfactants (in the form of new synthetic processes, the classical route of reacting acetylene with ketones had been discovered in the 1930s) and the fluorosurfactants. These brilliant developments have broadened the boundaries of colloid chemistry and enabled solving extremely demanding formulation problems.

However, their commercial relevance is still no match for the large volume "conventional" surfactants. This is in part because the high costs of the products may have discouraged their widespread evaluation but also because the applications in which their unique effects can be advantageously exploited, although quite large in number, consume comparatively small volumes of finished products and/or require only a low level of usage of the surface active agent.

In industrialised countries, surface active agents have been well established for many years and the surface active agents industry is presently at a standstill, to say the best. The chemistry, manufacturing and application technologies are established and well known. The low capital demand for a surfactants plant compared to other chemicals attracted investments and created a situation of endemic over-capacity in nearly saturated markets that only recently began to be re-absorbed. No matter how clever and creative the surfactants marketers are, there are physical limits to the number of laundry or personal washes that can be performed in a given time, whilst consumption in the industrial sector, generally following GDP trends do not benefit anymore from the quantum leap it had experienced earlier in the past century. Another limiting factor to the growth in the US and Europe is the change of the industry pattern that is progressively moving towards services at the expense of heavy industrial production. There is clearly a difference in the volume and type of surfactants that were traditionally used in a steel mill from the ones required to produce an electronic device. It might be argued that the demand for surfactants is simply shifting away from the industrialised to the developing countries, following the migration pattern of the industry and the rising consumption for personal and home care. This is correct, and the size and the growth of the global surfactants market in recent years were determined by emerging economies. The majority of consumption is of the classical all-purpose workhorses, while the more sophisticated effect-directed species, albeit growing, are far from the level reached in industrialised countries.

Nowadays, the industry of surfactants, its raw material producers, and its formulators is permeated by the concept of sustainability, which is producing important structural developments. Nonyl phenol ethoxylates are in retreat, even in low or non-regulated countries, replaced by fatty alcohol ethoxylates (mostly of natural origin) and fully sustainable anionic surfactants may challenge the absolute king of the surfactants: the linear alkylbenzene sulphonate.

BIBLIOGRAPHY

(1) Henshilwood, C.S., et al., "A 100000 Year Old Ochre-Processing Workshop Blombos Cave, South Africa" *Science* 334, pages 219–222 (2011)

Part I

Natural Surface Active Agents

1 Soap, the First Man-Made Surface Active Agent

Any attempt to reconstruct the historical development of soap and soapmaking meets with two major obstacles.

First, soap making is a craft often neglected in history. Subjects like battles, migrations, births, marriages, deaths, kings, queens' and nobles' genealogies, intrigues, edicts, natural catastrophes, religious confrontations, power struggles, agriculture, food, textiles waving and dyeing and so on and so forth are much more important to be consigned to history than the discovery and evolution of preparations for washing clothes or cleaning bodies.

Second, an archaeological search is crippled by the fact that soap is soluble in both organic solvents and water, and its remains are unlikely to survive. Once used, soap no longer forms part of a permanent record; thus, the probability of finding any residue is minimal (although very ancient remains were discovered, see the following discussion). The vessels and equipment for soap production would be indistinguishable from those used for other purposes, with perhaps the exception of ash leach barrels (and even for those, other possible uses can be envisaged).

The oldest reference to the production of soap by a purpose-specific chemical process was found on clay cylinders excavated at the site of ancient Babylon and containing a soap-like material. These are dated to around 2800 BCE. The inscriptions on the cylinders explain that fat was boiled with a slurry of ashes. It does not refer to the properties or the purpose of the composition.

The first report of the "detergent" properties of soap is engraved on a Sumerian clay tablet from around 2500 BCE found near the village of Tello in Mesopotamia that gives instructions for washing wool with "soap". It is perhaps surprising to us that soap appears to have been valued more for use with textiles than personal hygiene. In ancient times, however, colourful garments were particularly expensive and, as such, a symbol of wealth and status. Wool, as it comes from sheep, is coated with a layer of grease (lanolin) that interferes with the application of dye: washing away the grease enables uniform colouration and high colour yield, hence the recognition given to a product that did the job effectively.

By 2000 BCE, Alep, already a cross-point of Western and Eastern trade, had also developed a reputation in soap manufacturing thanks to the innovation of adding to olive oil a certain amount of laurel-berry oil. This imparted to the soap a soothing and insect-repellent effect, and the combination is still commercially exploited today in handcrafted formulations.

The Ebers Papyrus, a medical document dating back to the 18th dynasty of Egypt (more precisely to the time of Amenhotep I, about 1550 BCE) is a formulary of over 800 prescriptions and 700 drugs. Among these, there is a recipe combining animal and vegetable oils with an alkaline material called "Trona" collected in the Nile

DOI: 10.1201/9781003403869-2

Valley to produce a substance suitable to treat skin diseases and for personal cleaning. Other ancient Egyptian papyri mention soap-like substances used in the preparation of wool (1).

According to the Greek historian Herodotus, the Scythians living in what is modern Ukraine were using soap to wash their hair as early as the 7th century BCE.

It is possible that soap was used for personal hygiene and not only for hair washing, but there is no clear evidence to support this, apart from two mentions in the Old Testament: "For though thou wash thee wit nitre, and take thee much soap, yet thyne iniquity is marked before me, saith the Lord God" (2). There are doubts as to whether this is a reference to true soap or simply a lye made by mixing alkaline plants with water and oil, as reported in the book of Job: "The ancients made use of alkali mingled with oil instead of soap" (3). The Hebrew word that was translated as "soap" in the texts cited is *borith*, a derivative of *bor*, "purity", hence something which makes pure and may refer to any substance that cleanses, for example, the vegetable lye called in Arabic el-Qali (the mixture of crude sodium and potassium carbonate obtained by burning desert plants). Even today there are remote districts in Syria where cooking utensils, clothes and even the body are cleaned with ashes. It may as well have been a lathering paste of fuller's earth and impure soda mixed with wine, of the kind used by the Romans to clean clothes and pots and found in the ruins of Pompeii, as possibly suggested by a statement in the book of Malachi "He is like a refiner's fire, like a fuller's soap" (4) (note that in ancient times el-Qali was used as a flux in refining precious metals). In the deuterocanonical appendix to the book of Daniel, Susanna is using a substance described as "smegmata" for bathing in the orchard: "Dixit ergo puellis: afferte mihi oleum, et smegmata, et ostia pomarii claudite, ut laver" ([she] then said to the young: bring me oil, and smegmata and close the entrances to the orchard, such that I can wash [myself]). The word *smegmata* has its origin from the Greek *smegma/smechein*, "to wash off, clean", and probably indicates some sort of el-Qali which was mixed with oil, applied to the body and rinsed off.

Luther in his translation of the Bible used the word *Seife* to describe the same substances cited earlier. If this may have added to the confusion about their true nature and use, it proves at least that at the time soap was sufficiently known and recognised to be used in a work aimed at a large audience.

In ancient times, commercial routes had already been developed to an amazing extent. If no evidence of trans-continental trade with the Americas has been discovered so far (yet cocoa and tobacco leaves have been found in Egyptian mummies; how did they get there in the first place?), it is, however, a reality that the commercial links which bound the ancient world were both extensive and sophisticated. The Saharan caravans, the Arab dhows plying the Indian Ocean, the fleets of Chinese junks coasting East Asia and the most famous Silk Road formed a truly intercontinental trading system, and goods and technologies travelled thousands of miles to and from Scandinavia, Britain, Egypt, Crete, and the Middle East, where they met the ships and caravans from Mesopotamia, Arabia, India, and China.

In their journeys, traders are likely to have sailed and laid anchor along the coasts of Greece, southern France, the Iberian Peninsula, Italy and of the North Sea. No wonder, then, that soap-making technology spread from its birthplace to the whole of the Mediterranean basin, especially Greece because of the abundance of olive oil, southern France, where the saponification of olive oil was presumably a common practice in the Greek colonies of Provence and where any shortage of this raw material could be made good with the abundant supply of animal fat from the interior of the country and southern Italy, driven by the Greek settlements of Magna Graecia. In other Mediterranean countries, soap making did develop more slowly, probably because the limited demand from a scarce and unsophisticated population could be initially fulfilled by trade. If not from the Greeks, the Gauls may have gotten recipes for soap making from Egypt through Phoenician sailors as a description of soap and of its properties as we have seen was already recorded there 1500 years BCE. It is documented that soap was an article of trade for the Phoenicians as early as the 6th century BCE (5).

Soap manufacturing was undoubtedly practised in the Roman Empire, although it is still uncertain to what extent. According to a legend, soap got its name from Mount Sapo, a hill along the river Tiber where animals were sacrificed. Rain washed the mixture of fat and ashes down into the clays of the river's bank, which acquired the properties to wash clothes cleaner and with less effort (6). Other sources, however (the *Partridge Etymological Dictionary*), suggest that *sapo* is derived from the Germanic or Celtic language, the native reflex of which was "sebum".

In 1748, excavation work at Pompeii unveiled a workplace that, because of its design and the materials found, was beyond doubt a soap factory and shop, preserved remarkably intact over the centuries by the thick layer of lava and ashes deposited during the deadly eruption of the Vesuvius 79 CE (7).

Plinius the Elder (Caius Plinius Secundus, 23–79 CE) asserts in his *Historia Naturalis* (book 28, chapter 51) that the Gauls were the first to produce soft and hard soaps from goat fat and wood ashes (8). "Prodest et sapo, Galliarum hoc inventum rutilandis capillis. Fit ex sebo et cinere, optimus fagino et caprino, duobus modis, spissus ac liquidus, uterque apud Germanos maiore in uso viris quam feminis" (Soap is the invention of the Gauls, and this is used to redden the hairs. [It] is made from fat and ashes, the best [ingredients being] beech wood ash and goat fat, combined in two ways [to give] a thick or a fluid [product]. Among the Germans men use it more than women).

It is unlikely that the Gauls (or the Celts as reported by some authors) can be credited with the "invention" of soap, given the long history of development and the likely degree of sophistication achieved by soap making in the Middle East and Egypt and the extent of the commercial and technology exchanges across the Mediterranean basin and the North Sea. It is more likely indeed that the Gauls/Celts may have rediscovered a soap made from animal fat, as opposed to the vegetable sources probably in favour at the time in the Mediterranean countries. It is certain that animal fat was originally used by the early Middle East civilisations: Plinius the Elder express the view that the Phoenicians prepared soap from goat tallow and wood ash already in the 6th century BCE and may have disseminated the knowledge during their trips. This incidentally strongly supports the interpretation that the invention reported in the *Historia Naturalis* refers to the practice of colouring hair rather than to the discovery of a colouring substance (9).

This is the only mention of the word *sapo* in all of Pliny's work, and no mention is made of using the stuff for washing up. This supports the hypothesis that at the time, soap was only consumed for aesthetic purposes: it appears that the sophisticated Roman ladies used soap to get shinier, lighter colour hairs (a strongly basic soap has indeed a hair bleaching effect) and the Gauls and the Germans to both tint and wash their hair.

That hair aesthetics were important in ancient Rome is confirmed by the writing of Ovidius (Publius Ovidius Naso, 43 BCE–18 CE): "Femina canitiem germanis inficit herbis, et meliore vero quaeritur arte color" (The women colour their white hairs with German herbs, thus achieving an appearance better than the natural).

It is possible that the Germanic tribes may have contributed also to the spread among the Romans the use of soap for personal wash over their traditional bathing habits. Soap from *Germania* was highly praised in Rome in imperial times, and in the second century CE, the Greek Galen (129–201 CE), the personal physician of the emperor Marcus Aurelius (121–180 CE) recommended soap for both medicinal and cleansing purposes. Similar recommendations were made a few years later by Serenus Sammonicus, the physician of the emperor Settimius Severus (146–211 CE).

In 385 CE, Theodorus Priscianus recommended soap for washing hair and gave the first description of the "saponarius" or soap maker (10).

The disintegration of the Western Roman Empire, the weakening of the Eastern Empire after the disastrous Gothic wars and the later Moslem expansion into the Middle East, northern Africa, and Spain caused the disruption of the trading routes across the Mediterranean basin. They pushed the whole of Europe into the darkness of the Middle Ages. It was not to recover until the fragmented land-based feudal structure and the consequent close economic system faded away in front of the emerging centralised monarchies and the trading activities of a town-based class of merchants and bankers.

With the social, economic and cultural regression of the Middle Ages, bathing and soap fell from favour in Europe. However, despite the general social, cultural and technological regression and the disruption of communications soap making and some limited soap trading continued.

A guild of soap makers seems to have existed in Naples as far back as 599 CE (11). By the 7th century CE, soap manufacturing was an established craft in southern Europe, especially France, Spain and Italy, driven by the abundant availability of olive oil. Under the emperor Charlemagne (742–814 CE) soap making became a recognised application within the imperial domains. The *Capitulare de Villis vel Curtis Imperialibus* issued around 794 CE makes three specific references to soap:

- Chapter 43: "Ad genitia nostra, sicut institutum est, opera ad tempus dare faciant, id est linum, lanam, waisdo, vermiculo, warentia, pectinos laninas, cardones, **SAPONEM**, unctum, vascula vel reliqua minutia quae ibidem necessaria sunt" (. . ., i.e., linen, wool, woad [vegetable pigment], red dye, madder, carding implements, combs, **SOAP**, oil, container, and other small things that are needed there).
- Chapter 59: "Unusquisque iudex quando servierit per singulos dies dare faciat de cera libras III, de **SAPONE** sextaria VIII et super hoc at Andreae,

ubicumque cum familia nostra fuerimos, dare studeat de cera libras VI; similiter mediante quadragesima" (Every steward shall, when he is on service, give three pounds of wax and eight *sextaria* of **SOAP** each day; in addition, he shall be sure to give six pounds of wax on St Andrew's Day, wherever we may be with our people, and a similar amount in mid-Lent).

• Chapter 62: "Ut unusquisque iudex per singulos annos ex omni conlaboratione nostra quam cum bubus quos bubulci nostri servant, quid de mansis qui arare debent, . . ., quid de fenus, . . . quid de melle et cera, quid de uncto et siu vel **SAPONE**, quid de morato, . . . omnia seposita, distincta et ordinata ad nativitatem Domini nobis notum faciant, ut scire valeamus quid vel quantum de singulis rebus habeamus" (That each steward shall make an annual statement of all our income, from the oxen which our ploughmen keep, from the holdings which owe ploughing services, . . . from hay . . . from honey and wax, from oil, tallow, and **SOAP**. All these things they shall set out in order under separate headings and shall send the information to us at Christmas time, so that we may know the character and amount of our income from the various sources).

Contemporary to the *Capitularis de Villis* is the *Mappae Clavicula* (approximately translated as "The Little Key of Handicraft"), possibly originated in northern France or Germany by the early 9th century. It is a compendium of recipes related to knowledge of ancient Egypt, Greece and early alchemist texts. It has descriptions of the nature and preparations of minerals, herbs and chemicals; recipes for making glues, pigments and soap; and instructions for dyeing textiles. The two reported recipes for soap making read as described (12) below:

Soap Made from Olive Oil or Tallow

Spread well burnt ashes from good logs over woven wickerwork made of tiny withies, or on a thin-meshed strong sieve, and gently pour hot water over them so that it goes through drop by drop. Collect the lye in a clean pot underneath and strain it two or three times through the same ashes, so that the lye becomes strong and colored. This is the first lye of the soap maker. After it has clarified well let it cook, and when it has boiled for a long time and has begun to thicken, add enough oil and stir very well. Now, if you want to make the lye with lime, put a little good lime in it, but if you want it to be without lime, let the above-mentioned lye boil by itself until it is cooked down and reduced to thickness. Afterwards, allow to cool in a suitable place whatever has remained there of the lye or the watery stuff. This clarification is called the second lye of the soap maker. Afterwards, work [the soap] with a little spade for 2, 3 or 4 days, so that it coagulates well and is dewatered, and lay it aside for use. If you want to make [your soap] out of tallow the process will be the same, though instead of oil put in well-beaten beef tallow and add a little wheat flour according to your judgement, and let them cook to thickness, as was said above. Now put some salt in the second lye that I mentioned and cook it until it dries out, and this will be the afronitrum for soldering.

French Soap

Agitate with cold water two parts oak ashes with a third of oak [sic] lime. Afterwards when they are well stuck together, put the whole in a basket, strongly pressed down to make on top a place for water so that it does not run away. In this you will put cold water two or three times according to the amount consumed by the underlying ashes and the lime. Not quickly but on the following day, the water will drip down onto leaves of laurel or the like placed underneath, so that later it may flow off into another pot, and this is called the capitellum. Now, if you want to make soap, put in a second water following the first, and when that has run down, put in also a third, and it will be good until it becomes white. Afterwards, melt some tallow, strain it, and when it is strained and cleaned on top if necessary, boil it with the last water. When it becomes thick, put in some of the second water and also some of the first. Or else if you soak ground poplar berries for a day [in the mixed lye and tallow] and afterwards squeeze and discard them, the soap will be reddish and better. This is French soap and spaterenta, i.e., sharp.

In the preceding recipes, there is a reference to use of berries to colour the soap in red. Perhaps this explains Pliny's claim that the Gauls use it to colour their hair red.

The version from which this text was translated dates probably to the mid-12th century (possibly the beginning of the 13th century and attributed to Abelard of Bath) but does not appear in the earlier (incomplete) versions. It is likely that soap making at this time was already in the hands of guilds (traditionally extremely protective of their secrets), because only these two alchemic descriptions come down to us, and no other recipe is included in the "guides for good housewives" published towards the end of the period. What is historically proven is that guilds of soap makers existed in Augsburg (1324), Vienna (1331), Nuremberg (1357) and Ulm (1384) (13).

Outside the Mediterranean countries, soap making developed more slowly; for example, in England, it started only during the 12th century, although the Celts are credited by some sources with introducing soap to England around 1000 CE.

The Arabs first and later the Turks were the first societies to fully appreciate the real value of soap. The invasions of Spain by the former and of the Byzantine Empire by the latter and the Crusades contributed to the spread into Europe of the technology of soap making and the use of soap for cleaning purposes. Soap is repeatedly mentioned as a cleaning agent in Arabic writings from the 8th century onwards. Alchemists like Al-Razi described the concentration and purification of al-Qali, the ashes from barilla, (a salt-tolerant woody shrub that, until the 19th century, remained the principal source of soda ashes) and of oak wood to give pure potassium and sodium carbonate. He also gave some recipes for soap making and a description of a process for producing glycerine from olive oil (14). The process of craftsmen in Arab soap-making centres of Nablus, Damascus, Aleppo, and Sarmin used olive oil and al-Qali, and these came very much in vogue with the European nobility.

The Arabs are also credited for the discovery of hard soap. David of Antioch (Dawud al-Antaki) gave one recipe (Hassan and Hill, (14), pages 150–151):

Take one part of al-Qali, and half a part of lime. Grind them well and place them in a tank. Pour five times water and stir for two hours. The tank is provided with a plughole. When the stirring is stopped and the liquid becomes clear, the hole is opened. When the water is emptied plug the hole again and pour water and stir, then empty, and so on until no taste is left in the water. This being done while keeping each water separate from the other.

Then take from the pure oil ten times the quantity of the first water and place on a fire. When it boils feed it with the last water little by little. Then (add) the water before the last until at last you feed it with the first water. Then it becomes like dough. Here it is ladled out [and spread] on mats until it is partially dry. Then it is cut and placed on nura [slaked lime]. This is the finished product and there is no need to cool it or wash it with cold water while cooking. Some add salt to the al-Qali and lime in half the quantity of lime. Others add some starch just before cooking is over. The oil can be replaced by other oils and fats such as the oil of carthamus.

There is little doubt that this recipe refers to hard soap. Unfortunately, Hassan and Hill give no date for Dawud al-Antaki quotation, thus leaving some doubt as to the first production of hard soap.

Throughout the first millennium CE, soap was undoubtedly scarcely used and, as we have seen, only for cosmetic purposes in Roman times. Although claims of soap making at Pompeii in the 1st century continue to be made, Hoffman's analysis of this supposed soap in 1882 showed it to be nothing more than fuller's earth, a kind of alkaline clay described by Pliny for the laundering of clothes. It is also surprising that there is no mention of soap in the *Stockholm Papyrus*, a 2nd-century collection of dye recipes that gives instructions for washing wool with vinegar, ashes, and plants, without reference to anything resembling soap. Washing clothes with urine, plants, clay, and ashes was the common process across the Mediterranean basin.

We have seen that the 2nd-century physician Galen used the word *sapo* to describe a material made from goat tallow and ashes, which is good for medicinal and cleaning purposes. But while this is undoubtedly soap, his description occupies less than a few sentences in the thousands of pages which make up his complete works. Indeed, the cleaning habit of the Romans, other than bathing in the thermae, was to oil the body and scrap the dirty oil off with an instrument called a strigil.

It is likely that throughout the early Middle Age, the volumes of soap produced were so small that the production was done domestically, confined to or nearby the points of demand and that it remained a free craft until the 12th century. Indeed, we have seen earlier that Charlemagne demanded soap makers (as well as to gold- and silversmiths, bakers, etc.) to show up at places where the imperial court travelled (see the chapter 59 of the *Capitulare de Villis*). His son Ludwig the Pius in later times issued provisions for craftsmen travelling to marketplaces.

It remains unclear what the little soap produced was used for in those times. The reference in the *Capitulare de Villis* suggests that it was or may have been produced in relatively small communities, and there are reports of major production centres as of the 6th century. But what for? I can only offer hypotheses. Of the four possible uses,

- Cosmetics
- Personal hygiene
- Cloth washing
- Textile processing (wool scouring, dyeing)

the latter appear as the more likely. It is in later times that the use of soap for personal hygiene and cloth washing began to spread.

Along with the cultural and economic revival of the late Middle Ages, a soap industry started to develop around the 13th and 14th centuries. By 1200, the French city of Marseille became a major centre of soap manufacturing and remained so through the Middle Ages to the present. It was not long, however, before Venice, Genoa, and Bari in Italy and the region of Castile and the cities of Alicante, Malaga, and Seville in Spain established their own soap-making industry in competition with Marseille. Each of these areas had an abundant supply of olive oil and barilla (a fleshy plant whose ashes, rich in carbonates, were used to make lye).

In Britain, the origin and development of the soap business has been the source of considerable debate. The first question is, When did soap make its appearance in the country? This is attributed to French and Celtic influences as we saw earlier. References begin appearing in the literature from about 1000 CE, and in 1192, the monk Richard of Devizes refers to the number of soap makers in Bristol and the unpleasant smells which their activities produced. A century later, soap making was reported in Coventry. Other early centres of production included York and Hull. In London, a 15th-century "sopehouse" was reported in Bishopsgate, with other sites at Cheapside, where there existed Soper's Lane (later renamed Queen Street), and by the Thames at Blackfriars (15). The next question is, As soap making and trading were established, could a guild of soap makers be identified early in that period? According to D. R. Sherman, in "Domestic Lighting: Candles, Lamps and Torches in History" (16): "The Worshipful Company of Tallow Chandlers of London was constituted as a guild by letter patents from King Henry VI in 1422; however, the tallow chandlers were certainly organised long before they were granted their charter. Soap making was part of the purview of the tallow chandlers by 1509. In 1545, the London Authorities ordered that butchers 'that used to sell theyr tallowe to soapmakers [sic]' are 'not to sell yt in eny wyse to eny soapmakers upon the perylls that may fall thereon [sic]', thereby reserving tallow for candles. These quotes seem to suggest that there were 'soap makers' who were not in the chandler's guild and that politics were being played.

The other debate was on whether soap should be made from olive oil or tallow. Around the beginning of the 16th century, the French acquainted England with soap made with olive oil rather than animal fat. The 1609 book *Delights for Ladies* by Sir Hugh Plat mentions two recipes for removing stains. One refers to a "white, hard "soape [sic]" and the other to "castill sope [sic]". Castille soap is just a name for pure olive oil soap. It is white and hard and is made with the ashes of plants with high sodium content. Soap made with potash from hardwood (as done in England) does not get hard. It was sold in pots or tubs and was also called "black soap".

As a rule of thumb, olive, beef and mutton tallow make very hard soaps; pork fat makes a soft soap; and chicken fat may end up with a jelly soap that never gets hard.

Both tallow and olive oil were used in ancient times as well as in the Middle Ages. However, it is generally believed that tallow was reserved for candle making. Olive oil seems to have been the preferred fat for soap making, as apparently shown by the Bristol and London soap maker's guilds. Olive oil was reportedly known in northern

Europe by the 9th century. Oils like rapeseed and canola may have been sometimes allowed, but other oils would be confiscated when discovered (17).

At least in Britain, 1597 brings the first evidence of soap being made in individual households. Domestic soaps made from tallow were produced in Bristol, Coventry and London.

By the 17th century, the emergence of soap as a regular article of commerce did not escape the attention of those seeking to raise money from taxation. In 1622, King James I granted the monopoly of soap making to one company against the payment of a duty of 20,000 pounds sterling a year (the equivalent of 100,000 USD in today's money) for the production of 3000 tons of soap. This caused trouble, as some 20 other producers who had been excluded from the deal did not want to recognise the company's rights. The king then ordered that soap could not be sold unless it had been previously approved by the company. In 1632, Charles I granted letters of patent to the Society of Soapmakers of Westminster, granting them a 14-year monopoly of the production of certain types of soap in return for payment of 4 sterling pounds per ton (18). As the board of the manufacturing company included Catholics, the term *Popish Soap* (after the pope) was applied to this monopoly commodity. It was said by anti-Catholics to be particularly harmful to linen and washerwomen's hands (19).

Soaps made from olive oil were imported from Marseille, Venice and Castile; domestic soaps made from tallow were produced in Bristol, Coventry and London. The scale of the industry was increasing: in 1624, the Corporation of Soapmakers at Westminster was granted a royal patent to produce 5000 tons of soap per year (20).

In 1633, 16 soap makers were convicted for breach of the patent rights and were sentenced to fines of between 500 and 1500 sterling pounds and jailed for "as long as it pleased to his majesty". Two of them died in captivity, others were detained for 40 weeks. The rights of the company were renewed in 1635 against an additional payment of 2 pounds sterling per ton of soap. The company rights and assets were eventually sold in 1637 for a consideration of, respectively, 40,000- and 3000-pound sterling (21).

Taxation continued in various forms. During the Commonwealth period, it stood at 4 shillings per barrel. In the 18th and early 19th centuries, under a tax introduced under Queen Anne in 1712, the levy varied between 1 penny and 3 pennies per pound, the higher figure being equal to the total cost of production. All soap pans were required to be fitted with a padlock, of which the key was held by the exciseman. This official was required to attend each soap boiling, of which 12 hours' notice was required to be given.

The soap tax drove many soap makers out of the country. Many of them migrated to countries such as Ireland, where soap was exempt from tax. In addition, the tax resulted in widespread soap smuggling to avoid the tax.

The tax was unpopular with the poor as it tripled the price of the basic coarse soap. The soap tax resulted in making soap a luxury item and affordable only for the wealthy classes. It was eventually Gladstone, as Chancellor of the Exchequer, who, in a growing tide of Victorian concern about cleanliness, abandoned the soap duty in 1852, at an annual loss of sterling pounds 1,126,000 in tax revenue.

The soap industry in the UK flourished through the industrial and social initiatives of the Lever brothers. William Lever and his brother James entered the soap business in 1885 by buying a small soap works in Warrington. They associated with a Cumbrian chemist, William Hough Watson, who invented a process to produce soap using glycerin and vegetable oils, such as palm oil, rather than tallow. The resulting soap was a good, free-lathering soap, at first sold as "Honey Soap" and later as "Sunlight Soap". The enterprise was named Lever Brothers Company and become the first company that contributed to diffuse cleanliness in 19th-century England. Sunlight Soap was so successful that production reached 450 tons per week by 1888. In 1887, the Lever Brothers Company began looking for a new site on which to expand its soap-making business, which was at that time based in Warrington. The company bought about 25 hectares of flat, unused marshy land in Cheshire, south of the River Mersey. It had enough space for expansion and was conveniently located between the river and a railway line. The site became Port Sunlight, where William Lever built his works and a model village for his employees. Along with the construction of the village, he introduced welfare schemes and provided for the education and entertainment of his workforce, encouraging recreation and organisations which promoted art, literature, science and music. The purpose, he stated, was "to socialize and Christianize business relations and get back to that close family brotherhood that existed in the good old days of hand labor". William Lever claimed that Port Sunlight was an exercise in profit sharing, but rather than share profits directly, he invested them in the village. He said,

It would not do you much good if you send it down your throats in the form of bottles of whisky, bags of sweets, or fat geese at Christmas. On the other hand, if you leave the money with me, I shall use it to provide for you everything that makes life pleasant – nice houses, comfortable homes, and healthy recreation.

The Lever Brothers Company entered the United States market in 1895, with a small sales office in New York City. In 1898, it bought a soap manufacturer in Cambridge, Massachusetts, the Curtis Davis Company; it moved its US headquarters to Cambridge and started production there. By 1929, Lever Brothers employed 1000 workers in Cambridge, and 1400 nationwide, making it the third-largest soap manufacturer in the US.

In 1914, Lever Brothers took a majority shareholding in the A. & F. Pears Company, the producer of "Peers" translucent soap, and completed the takeover in 1920. Pears was the first mass-market translucent soap, produced by a special manufacturing process that included the neutralisation with potassium hydroxide, the use of methylated spirit and a long maturation time. It is still an important brand in the Unilever soap range.

In 1929, Lever Brothers merged with the Dutch margarine producer Margarine Unie to become Unilever. It is now the largest soap producer in the world.

By the 18th century in Europe, soap was a relatively common domestic item. The *Dictionaire Oeconomique* of 1758 describes:

SOAP, a Composition made of Oil of Olive, Lime, and the Ashes of the Herb Kali or Saltwort; the chief use of Soap is to wash and cleanse Linnen: There are two sorts thereof, which are distinguished by their Colours, viz. White and Black Soap.

Despite this progress, the consumption of soaps for personal hygiene was probably not high and must have bottomed out in the 17th and 18th centuries, and it is reported, for example, that the Palace of Versailles had only two bathtubs. One was assigned to the Marquise de Pompadour, who apparently used it to accommodate flowerpots. There is no doubt that the lack of personal cleanliness and the related unsanitary living conditions contributed heavily to the great plagues of the Middle Ages.

In those days, soap making was not free from risks: it is said that Louis XIV had three soapers executed for having made a bar that irritated his royal skin. In desperation, the surviving craftsmen got together and invented a method for producing a "mild" soap using a pouring and curing process that is reported to have taken a month to yield a single bar. The credibility of the story is doubtful; however, it seems to be a fact that the king felt uncomfortably sick after each of his (very rare) baths and may have passed a draconian sentence after a particularly bad experience.

As we have seen Marseille is credited with making fine, gentle soap from local olive oil already in the 13th century. In 1666, Pierre Rigat, a merchant from Lyon, offered to King Louis XIV to produce enough soap to satisfy the whole demand of France without having to import raw materials from abroad.

He was granted a 20-year privilege to set up factories for producing any type of soap in any part of the country as pleased him. The existing six or seven soap factories could continue their activity only under the condition that they should not expand their assets and that all their production was to be sold at a fixed price to Rigat, who had the ultimate right to commercialise. The patent proved, however, difficult to enforce and was withdrawn in 1669.

In response to widespread cheating in soap making, the Edict of Colbert of 1688 forbade the use of any fat other than olive oil and gave rather detailed rules for soap manufacturing and trading (including some kind of anti-trust measures) to be applied to the "Savon de Marseille". Violations of these rules, and of the one stipulating that only olive oil was allowed in the production process, could have resulted in confiscation of the goods or even expulsion from Provence.

The edict was modified in 1754 and 1760 with respect to the time of the year during which soap making was allowed and was eventually wiped out by the French Revolution (22).

Little is known of soap making in German-speaking countries, apart from the establishment of the first guilds and the biblical references reported earlier. Only the availability of synthetic soda and tropical oil caused soap making to transform from an artisan craft into an organised and structured industry around 1830–1840. Some of the early technologies used to produce soap from coconut, palm, and palm kernel oil, as well as for the bleaching of these oils, came to Germany from England through the contribution of Kendall and Watt.

In the US, soap making goes back to 1608, with the arrival of soap makers on the second ship from England to reach Jamestown, Virginia. Even though the idea of bathing was surely not what it is today, keeping oneself and the household clean was nevertheless a common practice. Soap was therefore a valuable item, as shown by the efforts of the settlers in the New World to ensure its supply. The *Talbot*, a ship chartered by the Massachusetts Bay Company to carry people and supplies from England to its colonies at Naumbeak (the Salem and Boston of today), lists among its cargo two firkins (wooden, hooped barrels with a volume of about 35 litres) of soap. John Winthrop, who was to become the first governor of the Massachusetts Bay Colony, when writing to his wife from Boston in 1630, listed soap as one of the items to be brought along on her travelling to the New World.

At about the same time, Colonial Americans, faced with a shortage of soap imported from England expanded the practice of home production, and soap making remained a household craft for many years.

The task was labour-intensive and met with frequent failures. The diary of Elizabeth Ranch Norton, a niece of President John Adams, tells us how excruciating soap making could be. On one occasion, Mrs. Norton had to make three batches of soap before she could get the barrel she needed for her family.

Soap making was a yearly or semi-annual event. Many homesteads and farms used a large amount of tallow and lard from animals butchered in the fall. Alternatively, women stored waste cooking grease and fire ashes all year long and made soap before the spring cleaning. The lye solution was prepared by trickling water through the ashes, and having lye of the correct strength was essential for the success of the production. A fresh egg or potato was the instrument used to test the strength of the lye: if the object floated with a specific amount of its surface above the lye solution (about a quarter of a 25-cent coin today), it was considered suitable for use. If it exposed a larger area, the concentration was too high and water should be added; if the area was smaller or the object dropped, the water should be eliminated by boiling or the lye solution had to be poured through a new batch of ashes. Solid fat had to be cleaned, boiled and skimmed to extract dirt and other debris. The lye was then stirred into the fat. The formation of a thick mass meant success; if separation occurred, reworking was necessary.

The difficulties in soap making arose from the ignorance of the chemical process involved and led to many superstitions: good soapmaking was associated, in turn, to the phase of the moon, the tides and the equipment used. A Pennsylvania Dutch recipe states that a sassafras stick was the only suitable instrument for stirring the soap and that stirring must be done always in the same direction.

Potash (the residue remaining after all the water has been driven off from the lye solution) and pearlash (made by backing potash to burn off carbon impurities) were important raw materials, not only for soapmaking but also for industrial processes, for example, the manufacturing of glass. They were valuable items of trade to England being one of the few items produced by the American colonies that could be exchanged directly for cash. They were equally highly valuable for England, and their export to ports and countries outside British control was prohibited. To protect such valuable trade items, the Massachusetts General Court in 1755 established a code for assaying and standardisation of potash and pearlash. The newborn United States also understood the value of potash and pearlash manufacture: the first patent awarded by the United States Patent Office was to Samuel Hopkins for his technique of preparing pearlash.

In parallel with home-making, soap making by traders, called soap boilers, had grown with the expansion of the colonies since the early Jamestown settlement. Since tallow was the main ingredient for both soap and candles, many tradesmen produced both and were also called chandlers. Christopher Gipson, who landed at Dorchester, Massachusetts, in 1630, and then later in 1649 was elected Surveyor of the Highways of the Town of Boston, was a soap boiler. Since then (as now) engaging in politics requires both money and influence, it may be concluded that soapmaking provided Mr. Gipson with both. Josiah Franklin, the father of Benjamin, was among the number of the 18th-century soap boilers. By the mid-18th century, professional soap makers like Tallow Chandlers and Soap Boilers began to collect stored fat from households in exchange for some soap.

Soap was distributed door to door before selling in general stores became common practice. Hard soap was cut off from large blocks and wrapped up for household consumption; soft soap was distributed in barrels.

During the American Civil War, soap had uses other than bathing., for example, in whiting preparations for accoutrements, as described in the *Military Dictionary* (23), or mixed with brandy to treat wounds to horses or to treat horses for urinary retention.

As often in history, war was the stimulus for innovation and creativity. When materials for making soap in the traditional way became scarce, people resorted to cottonseed oil (already developed as fat for human consumption and industrial uses in the early 1850s), sunflower seed oil, rosin and various saponin-rich concoctions. Prickly pear was discovered to harden soap or candles, replacing the hard-to-find kitchen salt.

In 1806, William Colgate started a soap-making business in New York called Colgate & Company. It began selling individual bars of uniform weight in 1830 and, in 1872, introduced Cashmere Bouquet, a perfumed soap. By 1850, soap making was one of the fastest-growing industries, driven by a broad availability of raw materials and the move towards higher standards of hygiene. This transformed soap from a luxury item to an everyday necessity.

About the time the Colgate Company introduced Cashmere Bouquet, William Procter and James Gamble, two entrepreneurial characters from Cincinnati (at the time the capital of the key pig tallow raw material and therefore nicknamed "Porkopolis-pig city"), started a soap and candle business, peddling their product house to house from a wheelbarrow. Within a few years, they had established a solid manufacturing basis and a distribution network along the Ohio River. Over this time, the Procter & Gamble Company had worked to develop a high-quality soap, comparable to the imported Castiles, at an affordable price.

In 1878, the formulation was completed (credit for its development goes to James Norris Gamble, the son of one of the founders) and branded "White Soap" for the market introduction. A production accident made the first fortune of White Soap: a large batch of the soap was mixing when the workman in charge went to lunch and left the machine running. On his return, he realised that air had been worked into the mixture but decided nevertheless to continue production and poured the soap into the frames, where it hardened and was subsequently cut, packaged and shipped. A few weeks later, letters started pouring into Procter & Gamble asking for more of "the soap that floated". A human mistake had become a differentiating selling point. But why was the product so popular? Well, some people used to bathe in the Ohio River, and the floating soap would never get lost. The following year, Harley Procter, the son of one of the founders, capped the technological success with a spice of marketing glamour when, inspired by a Bible reading at a Sunday church service, he branded the soap "Ivory", a name undoubtedly more appealing than the original "White Soap". The inspiring Psalm 45 reads: "All thy garments smell of myrrh and aloe and cassia, out of the ivory palaces whereby they have made thee glad".

In the last years of the 19th century, the advent of petroleum and of electricity made available large volumes of stearine hitherto used to make candles. This gave the idea to the Procter & Gamble entrepreneurs to convert the surplus into a flaked soap, which was easier to handle and use for dishwashing and laundry purposes than the hard bars hitherto available. In line with the biblical Ivory tradition, the product was named Ivory Flakes (later to become Ivory Snow) and was to be followed by Chipso, the first soap designed for washing machines. It was, however, not until 1926 that Procter & Gamble introduced a perfumed beauty soap (Camay), almost two generations later than Colgate's Cashmere Bouquet.

In the western United States, the B.J. Johnson Company was making a soap based exclusively on palm and olive oil and met with such success that the company was renamed after it: "Palmolive". Palmolive later merged with a Kansas soap manufacturer, Peet Brothers, to become Palmolive-Peet, which in 1928, joined the Colgate Company to become the Colgate-Palmolive-Peet Company. Meanwhile, UK-based Lever Brothers were sending some of their staff to the US to get a position in that market and introduced Lifeboy Soap in 1895.

The presence of many major players and the demand from a growing market transformed the soap industry into a multibillion-dollar business, with companies determined to achieve market leadership through technological innovation, that is, the laundry detergents introduced at the beginning of the last century and aggressive marketing and advertising.

BIBLIOGRAPHY

(1) ENI School – Energy and Environment "History of Detergents", Aug. 3, 2016).
(2) Jeremiah, 2, 22 as reported in The Holy Bible, Authorized version 1611
(3) Job, 9, 30, Jeremiah, 2, 22 as reported in The Holy Bible, Authorized version 1611
(4) Malachi, 3, 2, Jeremiah, 2, 22 as reported in The Holy Bible, Authorized version 1611
(5) Meyers, Drew, *Surfactants Science and Technology*, John Wiley & Sons, 2005, page 3
(6) Willcox, M., *Poucher's Perfumes, Cosmetics and Soaps*, Vol. 3, 9th ed., Blackie Academic (Chapman&Hall), London, 1993 pages 393 et seq
(7) Der Seifensieder, *Eine gruendliche Anleitung zur Fabrikation allen im Handel vorkommenden Riegel-, Schmier-, Textil-, und Toilettenseifen*, Hermann Fisher, ed., Voigt Verlag, Lipsia, B.F., 1904, page 2
(8) Pllinius, *Historia Naturalis*, Book 28, Chapter 59
(9) *Handbuch der Seifenfabrikation*, 3rd ed., Julius Springer Verlag, Berlin, 1906, page 2
(10) "The Manufacture of Soap" *The Australian Journal of Hospital Pharmacy* 4, pages 33–39 (1974)
(11) Stefan's Florilegium, www.florilegium.org

(12) *Mappae Clavicula: A Little Key to the World of Medieval Techniques*, English translation by Cyril Stanley Smith & John G. Hawthorne. Transactions of the American Philosophical Society, n.s. v.64, pt. 4 (1974)

(13) *Handbook der praktischen Seifen-Fabrikation*, A. Engelhardt, ed., Erster Band, A. Hartleben's Verlag, 1896, page. 3

(14) Al Hassan, H.Y., Hill, D.R., *Islamic Technology, an Illustrated History*, Cambridge University Press, Cambridge, pages 149–151, ISBN 0 521 26333 6

(15) *The Pharmaceutical Journal*, 1st December 1999

(16) *The Complete Anachronist*, vol. 68, Society for Creative Anachronism; Milpitas, Ca, 1993

(17) *Proceedings, Minutes and Enrolments of the Company of Soap Makers, 1562–1642*, Harold Evan Matthews, ed., Bristol: Printed for the Bristol Record Society (1940), Bristol Record Society's Publications, vol. 19

(18) *The Pharmaceutical Journal*, 1st December 1999

(19) Martz, Dorilyn Ellen, and College of William and Mary. Department of History, *Charles I and "Popish Soap": An Exercise in Factional Court Politics*, 2000

(20) *Soapmaking* in Cavern Chemistry, Dunn Library Reserves, http://cavernchemistry.com

(21) *Handbuch der Seifenfabrikation*, Julius Springer Verlag, Berlin, 1906, page 3

(22) *Handbuch der Seifenfabrikation*, Julius Springer Verlag, Berlin, 1906, pages 3–4

(23) *Military Dictionary Comprising Technical Definitions, Information on Raising and Keeping Troops, Actual Service, Including Make-shifts and Improved Material*, Scott Henry Lee, D. Van Norstrand Publisher, New York, 1861

2 Soap-Making Process and Raw Materials

2.1 OILS, FATS, AND FATTY ACIDS

Animal fat was probably the first raw material in Mesopotamia before olive oil started to be used on the eastern Mediterranean shores.

Animal fats and olive oil were for centuries the source of fatty acids, and the saponification process remained empirical.

At end of the 18th century, the French chemist Chevreul (1786–1841) published the results of his research on the structure of oils and fats, which provided the scientific basis for the understanding and control of the saponification process. The possibility thus opened to use different sources of lipids for soap making, for example, cottonseed oil, sunflower seed oil, and rosin. The cost reduction from the improved manufacturing techniques, the recovery of the glycerine by-product, and the broader availability of soda (see the "Alkali" section) all combined to give a new impetus to the soap industry.

Natural fatty acids from fats and oils represented the most economical and viable route to raw materials for the manufacture of soaps. In Germany, however, the national plans for economic self-sufficiency led to the development of synthetic raw materials. The most important process involved the direct air oxidation of hydrocarbons (usually paraffin wax or products from the Fischer–Tropsch synthesis) at 100–180°C in the presence of a catalyst like potassium permanganate or other manganese compounds. The resulting complex mixture of oxygenated compounds (alcohols, ketones, acids) was then purified to separate the acids from the unsaponifiable matter. The process used industrially in Germany during World War II is described for example by Sheely (24). An excellent review paper with complete references on paraffin wax oxidation was produced by L. Mannes (25).

Synthetic fatty acids could be also produced by the oxidation of aldehydes obtained from olefins, carbon monoxide, and hydrogen in the presence of a catalyst (26), but this route does not seem to have had any commercial consequence.

2.2 ALKALI

From the beginning of the Christian era to the early 19th century, many spectacular inventions and discoveries took place: mathematics evolved from the impractical Roman numerals to the Arabic system to reach with Leibniz the sophistication of the infinitesimal analysis, the Americas and Australia appeared on the maps of the world, steam engines replaced water or animals power to operate factories, the geocentric theory was swept away by the Copernican revolution, Newton gave the first explanation of gravity as the mutual attraction of masses and anticipated the deflection of

DOI: 10.1201/9781003403869-3

light by gravitational fields, and Galileo set the foundation of modern science. Yet the description of a soap factory and of the soap-making process in the encyclopaedia of Diderot and D'Alembert bears more than casual reminiscences of what it looked like 18 centuries before.

The real progress in soap making started with the Leblanc process (1787) and, later, with the improved Solvay process to produce sodium carbonate. This eliminated the dependence on the scarce, low-purity, and comparatively more expensive alkali carbonates hitherto produced from wood ashes and enabled the large-scale saponification of cheap vegetable oils and animal fats. Sodium also gives a harder soap than potassium (of which wood ashes are particularly rich) that is easier to handle and use.

Prior to the discovery of Leblanc, the Benedictine monk Malherbe in 1777 used sodium sulphate as starting material to produce sodium carbonate, a process which was improved by de la Methrie in 1789. A few years earlier, however, Nicholas Leblanc (1742–1806), attracted by the price of 2400 livres promised by the French Academy of Science in 1775 for an effective process to produce soda, had worked on a different synthetic route based on sodium chloride. The process was fully developed by 1787. In 1790, Leblanc associated with his master, the Duke of Orleans, his treasurer and an assistant in chemistry at the College de France established a soda factory near St. Denis (Paris), which was confiscated at the onset of the French Revolution. In 1791, he obtained a patent on his process but forfeited it in response to the call of the Comité de Santé Publique to sacrifice personal interests for the benefit of the fatherland. Leblanc spent the rest of his life vainly attempting to get compensation for his discovery from the new establishment and, deeply depressed, killed himself in 1806 (27).

In the Leblanc process, sodium chloride is reacted with sulphuric acid (obtained at the time by distillation of hydrated ferrous sulphate) to produce sodium sulphate and hydrochloric acid. The sulphate is then roasted with limestone and coal, and the resulting mixture of sodium carbonate and calcium sulphide is leached with water to recover the sodium carbonate. The Leblanc process was expensive and caused significant pollution; however, it was practised at an industrial scale until 1916–1917.

The Solvay process, much more attractive economically and environmentally, was performed industrially for the first time in 1863 and, with some variations, is still in use today. The reaction

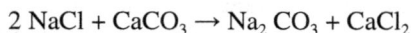

$$2 \text{ NaCl} + \text{CaCO}_3 \rightarrow \text{Na}_2\text{CO}_3 + \text{CaCl}_2$$

takes place in a series of steps.

(1) Sodium chloride is mixed with ammonia to give an ammoniated brine.
(2) The ammoniated brine in the presence of carbon dioxide (CO_2) gives sodium bicarbonate and ammonium chloride (sent to step 5).
(3) $2\text{NaHCO}_3 \rightarrow \text{Na}_2\text{CO}_3 + \text{CO}_2 + \text{H}_2\text{O}$
(4) $\text{CaCO}_3 \rightarrow \text{CaO} + \text{CO}_2$
(5) $\text{CaO} + 2 \text{ NH}_4\text{Cl} \rightarrow 2 \text{ NH}_3 + \text{CaCl}_2 + \text{H}_2\text{O}$

Ammonia is recycled to step 1, and CaCl_2 is sent to waste.

Other processes to produce sodium carbonate have been developed in more recent years by Asahi Glass, Huels and Akzo, but their relevance to the soap industry is less important and is not reviewed here.

BIBLIOGRAPHY

(24)   PB 2422, Office of Technical Services, Dept. of Commerce, Washington, DC.
(25)   Die Chemie, 51, Nr 1 and 2, 16, Jan. 8 (1944)
(26)   Ger. Pat. 734,219
(27)   Seifenfabrikation, 1906, page 5

3 Later Soap Developments

By the late 19th century, people had the choice of bar soap, soft soap, and powdered soap for their laundry. But soap suffers from a major drawback, which was the obstacle to a wider use. The fatty acids form insoluble salts with calcium and magnesium ions dissolved in water and form a curd-like precipitate that settles on the garments being washed.

The laundry-washing power of powdered soap was enhanced by combining it with other ingredients, and on June 6, 1907, the first "self-acting, heavy-duty" detergent was introduced by Henkel under the "Persil" trademark. Persil contained soap, sodium carbonate, and sodium silicate but, more important, a bleaching agent, and sodium perborate. The combination of detergency (from the soap surfactant and the alkalinity) and bleaching was an absolute novelty: the oxygen released by the perborate replaced the work that had been previously done by the housewife with washboard, scrubbing board, and brush.

On its own, this represented a step-change, although it was not a complete solution. The use of carbonate and silicate sequestrants reduces the precipitation by removing in part the multivalent ions, but some precipitation still occurs, and soap deposits gradually build up on fabrics, causing bad odours, the deterioration of colours, and the degradation of the clothes. Other disadvantages of soap are its limited cleaning power and deterioration during storage.

Synthetic detergents with higher cleaning power and more efficient sequestrants (polyphosphates) had to find their way in detergent formulations.

The demand for more effective washing products and practices was fostered by major social changes caused by the First and Second World Wars.

Before the World Wars, and especially the first one, few women worked outside their homes, and many households, even those of relatively modest means, employed domestic staff to cook and clean. After World War II, many women who had worked outside continued to do so, and fewer and fewer households could afford to pay for domestic help. There was then less time to do laundry, and the main driver of convenience fuelled the development of washing machines and better-performing detergent formulations suitable for the machine-washing process or facilitating hand-washing.

This was the last nail in the coffin for soap as the agent for linen and garment cleaning. More sophisticated formulations with appealing foam profiles, mildness, skin feeling, and scent have, however, allowed it to retain a significant presence in the personal care market in industrialised countries. In developing economies, soap is still a major player in hand laundry washing.

DOI: 10.1201/9781003403869-4

4 Naturally Occurring Surface Active Agents

4.1 INTRODUCTION

Many people understand the features and properties of soap, and it is therefore comparatively easy to find ancient references to its manufacturing and use. There are, however, several naturally derived substances with amphipathic structure and surface or interfacial properties that have been in use for as long as, if not earlier than, the fatty acid salts discussed earlier. Polysaccharides from acacia trees (e.g., *Acacia nilotica*, *Acacia Senegal*, and *Vachellia* (*Acacia*) *seyall*, commonly referred to as arabic gum) were used in the preparation of inks by the ancient Egyptians. By referencing this product and its effect, modern researchers have used arabic gum to separate the agglomerates of carbon nanotubes into individual fibres. A review of the industrial uses of arabic gum has been produced by Egbal and Amine (28).

Nature produces other subtances with interesting surface and interface properties, the most important are reported in the following subchapters.

4.2 SAPONINS

In ancient societies, and even today in many regions of the world, people clean their clothes by beating wet textiles on rocks near a stream. The mechanical agitation facilitates the removal of solid soil, and the water dissolves the hydrophilic stains composed of, for example, sugar, salt, starch, and certain dyes. Oily soil is, however, hard to remove in this way since fatty substances do not dissolve in water and remain attached to the fabric. The leaves, seeds, barks, and roots of several plants, are rich in surface active substances, and these have been probably used in the early days of civilisation for removing such soil.

Saponins are glycosides occurring primarily not only in plants (the presence in more than 100 families has been reported) but also in starfish (*Asteroidean*) and sea cucumbers (*Holothuroidea*). Properties generally shared by this group of natural products are surfactant activity, haemolytic action, steroid-complexing ability, and biocidal capability. The characteristic soapy lather formed when saponins are agitated in water is at the origin of the name. The basic chemical structure is an aglycone (steroidal or triterpenoid) linked to one or more sugar chains (hexoses, pentoses).

The use of saponins as cleaning agents probably predates recorded history. Today, these natural and sustainable surfactants have commercial applications such as ore separation, foaming agents in soft drinks, photographic emulsions, hair care formulations, medical and veterinary preparations, mild cleansing agents, and anti-dandruff shampoos.

Since antiquity, *Saponaria mukorossi* has been used in eastern Asia and the Himalayas as a detergent for shawls and silk and by jewellers for cleaning silver.

DOI: 10.1201/9781003403869-5

Saponaria officinalis, as Linnaeus described that "Bouncing bet – the Soap Weed" in his *Species Plantarum* (1753), was known to the Magi as "Chalyriton", to the Egyptians as "Oeno", and to the Africans as "Syris" (29). The Greeks identified the plant with the popular name "Catharsis", which meant, and still means, "a Cleanser or Purifier", perhaps referring to its use as soap and its purgative effect when used medicinally (30). The Romans called it "Radix" (i.e., root) or "Herba Lanaria" (i.e., wool herb), and already 2000 years ago, clothiers used a decoction of *Saponaria officinalis* to clean wool before weaving it into cloth. Its use was common in France and Germany already in the 10th century (where it remained in use until the 19th century) (31) and presumably across other countries in Europe in the late Middle Ages.

In England, the plant was called "bouncing bet". The name arose in Elizabethan times when it became widely used by working-class Britons who could not afford the exorbitant price of soap and used it to scrub their wood and pewter dishes and for their laundry. The prefix "bouncing", according to Haughton (32), may originate from the flower shape: "The inflated calix and scalloped petals of the flower suggested the rear view of a laundress, her numerous petticoats pinned-up, and the wide ruff at her neck bobbing about as she scrubbed the clothes". As far as "bet" is concerned, it may be speculated that the hypothetical washerwoman was named (perhaps ungraciously) after Queen Elizabeth.

The plant was also named "fuller's herb" in reference to its use in washing woven wool cloths because it not only cleaned the cloth but also caused it to shrink, pulling the strands tight or "full". When the Industrial Revolution enabled the large-scale commercial manufacture of cloths, textile industrialists cultivated fields of the "weed" as an inexpensive detergent (33). Bouncing bet was probably brought over to the Americas as part of the garden supply of the early settlers. Its cultivation boomed in the early days of the textile industry, and its fortune continued on the back of thrifty settlers who used it for any kind of cleaning, from fabrics to pottery as well as in balms, sheep dips, let alone to give beer a foamy head (e.g., among Pennsylvania Dutch) and to treat psoriasis, eczema, acne, and boils.

Common names in use today (latherwort, soaproot, scourwort) remember the plant's common use as soap. The suffix *wort* is probably the Anglicised version of an old North European word for plants or herbs; it has roots in the Anglo-Saxon *wyrt*, the Old Norse *urt*, and the Old High German *wurz* (34). It may even be related to the ancient Greek and Roman words for "root" (35). The word is obsolete by itself, but it survives as a suffix in many plants' common names. The prefix usually describes something a plant resembles (spiderwort), cures (liverwort), or provides (soapwort).

Today, the most exploited source of surface active saponins is perhaps the inner bark or the biomass of *Quillaja Saponaria*, Molina, a tree native of Chile, where it is used for washing clothes; cleaning delicate ribbons, garments, and wool; washing hairs; removing greasy spots; and disinfecting wounds. It avoids moth attacks on cloth, an indication of interesting natural insecticide properties. It has been used for over 100 years to produce triterpenoid saponin-rich extracts. Quillaja extracts are approved for human consumption in the US, the EC, and Japan, and they are used as a foaming agent in beverages (e.g., root beer), the production of slush-type drinks, blood analysis, and photographic emulsions. Purified saponins are also used more and more as vaccine adjuvants due to their immunological properties.

4.3   BILE ACIDS

A surface active agent not used for practical purposes yet of fundamental importance for human and animal metabolism, bile acids make up about 6–7% of bile.

Bile acids are converted to their sodium salts in the alkaline pH of the intestine and emulsify the food fats, thus facilitating their digestion.

Cholic acid exists conjugated with glycine as glycocholic acid and with taurine as taurocholic acid. Other bile acids present in smaller amounts have a similar carbon backbone.

4.4   PHOSPHOLIPIDS

Phosphatidyl choline, lecithin, lysolecitin, phosphatidyl ethanolamine, and phosphatidyl inositol are widely found in biological membranes. They are used as emulsifiers for intravenous fat emulsions and anaesthetic emulsions and for producing liposomes or vesicles for drug delivery.

Phospholipids play an important role in lung functions. The surface active material in the alveolar lining of the lung is a mixture of phospholipids, neutral lipids, and proteins. The lowering of the surface tension by this surfactant system and the surface elasticity of the surface layers assist alveolar expansion and contraction.

They are used in personal care for the formulations of liposomes for actives delivery and in pharmacy and veterinary sciences as emulsifiers for intravenous fat emulsions and anaesthetic emulsions and to produce liposomes or vesicles for drug delivery.

BIBLIOGRAPHY

(28)   Egbal, D., Amina, A., "Utilisation of Gum Arabic for Industries and Human Health" *American Journal of Applied Science* 10(10), pages 1270–1279, 2013

(29)   Gunther, R.T., *The Greek Herbal of Dioscorides*, ed. Hafner, Publ. Co., Inc., New York, 1933

(30)   Jaeger, E.C., *A Source-book of Biological Names and Terms*, 2nd ed., Charles E. Thomas, Springfield, IL, 1944

(31)   *100 Years of Revolutionary Research at Henkel*, page 38

(32)   Houghton, C.S., *Green Immigrants*, Harcourt Brace Jovanovich, Inc., New York, 1978

(33)   Georgia, A.E., *A Manual of Weeds*, The Macmillan Company, New York, 1942

(34)   Durant, M., *Who Named the Daisy? Who Named the Rose?*, Dodd, Mead & Company, New York, 1976

(35)   Simpson, J.A., Weiner, E.S.C., *The Oxford English Dictionary*, 2nd ed., Clarendon Press, Oxford, 1989

Part II

The Advent of Synthetic Surface Active Agents

INTRODUCTION

Although the Nekal naphthalene sulphonates are sometimes credited with being the first synthetic detergents, the reaction of sulphuric acid on oils and fatty acids can be traced to the end of the 18th century. The textile industry and the need for effective mordants (dye fixatives) stimulated researchers to investigate new domains and, probably driven by the easy access to raw materials commonly used in soap making, to study the modifications of oils, fats, and fatty acids caused by the reaction with strong inorganic acids like sulphonic and chlorosulphonic acid. Starting from 1790, this process continued throughout the 19th and early 20th centuries when the natural hydrophobic moieties began to be replaced with those of petrochemical origin.

In 1831, Fremy (a Gay-Lussac pupil) studied the action of sulphuric acid on olive oil, oleic acid, and almond oil. These products began to be increasingly used shortly thereafter in the textile industry. They can rightly claim the honour of being the first synthetic surface active agents, as they involved the addition of hydrophilic moieties to a hydrophobic substrate, as opposed to the oils and fats splitting and neutralizing the hitherto used soaps.

In 1925, a fundamental development in surface active agents started, with the introduction of Nekal, an isopropyl naphthalene sulphonate. This was the beginning of an extraordinary expansion of the surface agent industry, eventually resulting in the appearance of as many as 40 man-made surfactant classes, for a worldwide market of about 18 million tons (100% active) in 2018.

DOI: 10.1201/9781003403869-6

The converging forces that combined to ignite this innovative process were:

- The demand for easier and more effective laundering
- The ever-increasing hygiene standards
- The emergence of new industrial processes demanding higher-performing surfactants than soaps
- The advances in synthetic chemistry
- The availably of new raw materials, for example, olefines, alkyl benzene, ethylene oxide, and fatty alcohols

5 Amphoteric Surfactants

5.1 INTRODUCTION

Amphoteric surfactants carry simultaneously an anionic and a cationic hydrophilic group. The cationic part is a fatty amine or a fatty quaternary ammonium while the anionic moiety is a carboxylate, sulphonate, and phosphate.

There are two types of amphoteric surfactants, one which is pH-sensitive and another which maintains a cationic character at all pH ranges. The aqueous solution of the former type has different behaviour depending on the pH: in an alkaline medium, it exhibits anionic surfactant properties; in acidic conditions, it exhibits cationic surfactant properties. At the balance point of the cationic and anionic types (isoelectric point), it behaves as a nonionic surfactant.

Among the first amphoteric surfactants known are the higher alkyl taurines, synthesised by Rumpf by heating higher aliphatic primary amines with sodium bromoethane sulphonate (36).

Carboxy acid amphoterics are described in US Pat 2,206,249 (37).

Until the end of the 1940s and the beginning of the 1950s, amphoteric surfactants were regarded as curiosities: in 1949, Schwartz (38) wrote that

None of the ampholytic surface active agents have, to the writer knowledge, achieved any great commercial importance. It might be expected that at least some members of the group should exhibit unusual properties that could not be achieved by anionic, cationic, or non-ionic agents. The small amount of study which has been devoted to them has evidently not brought such properties to light as yet. A step-change in the commercial exploitation of amphoteric surfactants came in the 1960s with the introduction in the market by Th. Goldschmidt of the coco amido propyl betaine (CAPB) (39). This amphoteric has eclipsed the success of other secondary surfactants, because of an excellent toxicological profile, superior mildness, and complete biodegradability. Its isoelectric point is close to the pH of the skin (around 6.5), which produces a pleasant soft-skin feel (40).

Following this initial success, amphoteric surfactants have undergone rapid development in recent years, because of their low toxicity and low skin and eye irritation, as well as their excellent biodegradability; resistance to hard water; emulsifying, dispersing, wetting, foaming, and antistatic properties; and compatibility with all other types of surfactants. The main markets are in personal care (hair and body shampoos), followed by hand dishwashing (because of their mildness/foaming), and vehicle cleaners (because of rust inhibition).

In hair care formulations, amphoteric surfactants are used in combination with fatty alcohol sulphates to improve their solubility, reduce their irritation, and increase the viscosity and foam level and stability.

DOI: 10.1201/9781003403869-7

5.2   AMPHOTERIC SURFACTANTS PRODUCTION PROCESS AND RAW MATERIALS

Amphoteric beta-amino propionic acids are prepared by reacting, first, a primary fatty amine with methyl acrylate. After hydrolysis, the β-alkylamino-propionic acid is formed (41), (42).

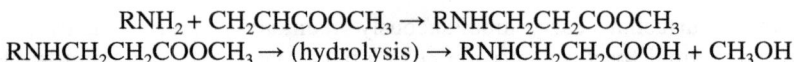

$$RNH_2 + CH_2CHCOOCH_3 \rightarrow RNHCH_2CH_2COOCH_3$$
$$RNHCH_2CH_2COOCH_3 \rightarrow (hydrolysis) \rightarrow RNHCH_2CH_2COOH + CH_3OH$$

Alkyl betaines are produced by reacting an alkyl dimethylamine with sodium chloroacetate (43).

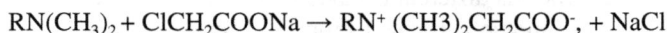

$$RN(CH_3)_2 + ClCH_2COONa \rightarrow RN^+ (CH3)_2CH_2COO^-, + NaCl$$

Mono- and dicarboxymethyl fatty amines can be produced from higher fatty primary amines by the formaldehyde–cyanide synthesis (44, 45).

$$RNH_2 + CH_2O + HCN \rightarrow RNHCH_2CN \rightarrow hydrolysis \rightarrow RNHCH_2COOH$$

Surface active amino carboxylic acids of this general type have also been produced by reducing the Schiff's base formed between a higher aldehyde or ketone and a lower amino acid (46).

An interesting type is the amido propyl betaines, developed commercially by Th. Goldschmidt (now Evonik) in the early 1960s. These are produced in two steps: first 3-dimethylamino propyl amine (DMAPA) is condensed with a fatty acid (C8—C18) or its methyl ester or a natural oil/fat to produce an alkyl amido propyl amine. During the reaction, the DMAPA is partly distilled off along with water or methanol, and it is isolated, purified, and recycled. In the second step, the amido amine is carboxymethylated with chloroacetic acid or sodium salt. In the final product (typically a water solution with 28–30% active matter), NaCl is present in high concentrations (up to 5%) (47). Apart from the reaction with the tertiary nitrogen chloroacetic acid may hydrolise to glycolic acid in a side reaction and is present in concentrations between 0.1–1.0%. A recent process patent (48) describes a method to increase the active content at 45%-plus by reducing the level of NaCl to below 2% and the pH to below 4.

BIBLIOGRAPHY

(36)   Rumpf, Compt. Rend. 212, 83 (1941)
(37)   US Pat 2,206,249
(38)   Schwartz, A.M., Perry, J.W., *Surface Active Agents, Their Chemistry and Technology*, Interscience Publishers, Inc., New York (1949)
(39)   Ger. Pat. 1,062,392 (1959)
(40)   Herrwerth, S., et al., "Highly Concentrated Cocamidopropylbetaine – The Latest Developments for Improved Sustainability and Enhanced Skin Care" *Tenside Surfactants Detergents* 45, pages 304–308 (2008)

(41) D.L. Anderson and A.J. Freeman, J. Cosmetics Chemists 8, 227 (1957)
(42) R.C. Freese, US Pat. 2,810,752 (1957) to General Mills Inc.
(43) Houben-Weil, *Die Methoden der Organischen Chemie, II, III*, Thieme, Stuttgart, p. 630 (1969)
(44) Brit. Pat. 460,372
(45) French Pat. 793,473
(46) Brit. Pat. 518,656
(47) Eur. Pat. 0 020 907 B1 (1982)

6 Anionic Surfactants

6.1 INTRODUCTION

It is difficult to conceive of something more peaceful than a housewife of the early 20th century busy washing her laundry, yet the surfactants industry owes much to the developments driven by the war economy in the First and Second World Wars.

Towards the end of World War I, the acute shortage of oils and fats in the Central Empires encouraged researchers to find alternative waste or non-strategic raw materials for producing surface active agents.

In later years, the search for substitutes for soap used in the production of styrene-butadiene rubber (SBR) by the emulsion polymerisation process led to the synthesis of the first surface active ethylene oxide derivatives.

If military and strategic considerations prompted the development of these new classes of surfactants, the technological advantages over soap ensured their commercial success, in particular the ease of use, the consistency of performance, the reduced sensitivity to water hardness, and the better surface and interfacial properties. The discovery that combinations of anionic surfactants and inorganic phosphates could provide better soil peptisation, hence cleaning, than soaps started a displacement process that was eventually going to relegate soaps to a secondary position in domestic and industrial detergency.

6.2 SULPHATED OILS AND FATTY ACIDS

Many technological developments are products of curiosity, combined with the courage of experiment and supported by the availability of materials easy to access or handle. It was probably a mixture of these ingredients more than anything else that lead Papillon in 1790 to use a mixture of olive oil and sulphuric acid as a mordant for textile dyeing (Depierre, Traite de la Teinture et de l'Impression), and it is perhaps this early empirical approach that led Fremy in 1831 to study in a systematic way the action of sulphuric acid on olive and almond oil and oleic acid.

The industrial use of sulphated oils dates from 1834. In his book *Chemie des Couleurs*, Runge reports the use of sulpholeate to replace "huile tournante" (rancid castor oil) in the dyeing with madder. Sulpholeates were used in 1860 by Gros, Roman, Maruzeau and Company in Wesselingen for printing fabrics with aniline dyes (49). and sulpholeic acid was recommended in 1864 for printing with aniline black.

The manufacture of "red turkey oil" from castor oil dates to 1875 (50) the use of sulphoricinates in the tanning industry in 1890 is documented in the work of Procter (51).

In 1896, Julius Stockhausen patented a process for the manufacture of jelly-sulphated castor oil that had increased stability towards salts of lime and magnesium, prevented discolouration during dyeing, and imparted a shiny appearance and a softness

DOI: 10.1201/9781003403869-8

to the fabrics (52). Further contributions to the chemistry of sulphonated oils came from the works of Grün (53) and Grün and Waldenburg (54).

Between 1911 and 1925, several articles on the treatment of vegetable and animal oils with sulphuric acid were published, as reported by Hildtitch (55). Van der Werth and Muller in their "Neuere Sulphonierung Verfahren zur Herstellung von Dispergier Netz und Waschmittel" report that more than 800 patents were granted for the sulphonation of oils, fatty acids, aromatic hydrocarbons, and derivatives.

In an article published in 1954, J.P. Sisley describes the reactions that take place during the reaction of sulphuric acid on saturated, unsaturated, and unsaturated hydroxy fatty acids (56).

6.3   SULPHATED ALKYL ESTERS

The esters of fatty acids with low-molecular-weight alcohols can be sulphated to yield surface active agents that, compared to the sulphated glycerides, have a greater foaming and wetting power. They were found particularly useful in the textile industry because of their lubricant effect and the high wetting and rewetting power, which is maintained in hot solutions of dilute caustic soda (such as are used in kier boiling) and in some of the continuous bleaching processes.

Propyl, butyl, and amyl esters of both ricinoleic and oleic acids were used as raw materials for the sulphation, with butyl oleate being probably one of the most commonly used (57, 58, 59). This type of product was introduced in Germany under the trade name Avirol AHX. It was produced in the late 1940s in the US, for example, as Surfax W.O. (Houghton), Phi-Sol (Onyx Oil and Chemical Co.), and Parapon (Arkansas Co.).

6.4   SULPHATED GLYCEROL ESTERS

Sulphated glycerol esters enjoyed considerable popularity in the 1940s and were commercialised in significant volumes by the Colgate-Palmolive-Peet Co. under the trade names of Arctic Syntex M, Arctic Syntex L, and Vel.

The mixed fatty acid/sulphuric acid esters of glycerol were studied first in 1909 (60), but their practical value as surface active agents was not appreciated at that time.

The real drive for the commercialisation of these products was their disclosure in two broad US patents (61).

The Syntex type of products can be made by both preparing a nominal monoglyceride first and then sulphating it or effecting simultaneously the esterification of glycerol with the fatty acid and sulphuric acid. The fatty monoglyceride can be prepared by the direct reaction of a fatty acid with an excess of glycerol or by treating a fatty acid triglyceride with an excess of glycerol. In both cases, an equilibrium mixture of mono-, di-, and triglyceride is obtained, which is dependent on the conditions of the glycerolysis. A third route (and the most advantageous one) is that practised by I.G. Farbenindustrie A.F. (and disclosed in the two Ger Pats. cited earlier) which consists in reacting simultaneously a fatty triglyceride, glycerol, and sulphuric acid or a fatty acid, glycerol, and sulphuric acid.

Glycerol ester sulphates had a brief moment of popularity when the (at the time) Colgate-Palmolive-Peet Company used in them in its formulation, but despite of this, the simplicity of the process and the cheapness of the raw materials, the Syntex type of surface active agents did not survive the competition from other sulphates/sulphonates. One of the possible reasons is their hydrolytic instability that prevented the spray drying of slurries (for many years, the only industrial route to powder detergent formulations) and the insufficient control of the residual impurities and the consistency of the products.

6.5   SULPHATED AMIDES

Fatty acids can be readily converted to amides by reaction with ammonia or primary or secondary amines. The amides can be sulphated and yield products with similar general properties as the sulphated esters (63, 64, 65).

The most widely used product in this class is the sulphated monoethanolamide of coconut fatty acid. The synthetic process involved the formation of the amide by direct amidification at 170–180 °C until reaching a low and constant acid number. The amide was then sulphated with sulphuric acid at 30 °C or less or with oleum or chlorosulphonic acid (66). In an alternative process, a fatty acid chloride was condensed with the pre-formed sulphuric ester of monoethanolamine in an alkaline solution.

The reaction of fatty acid and monoethanolamine also results in the formation of an amino ester, which cannot be sulphated, and remains in the product, detracting considerably from its wetting and foaming properties. The formation of an ester can be reduced if a fatty acid chloride is used (67, 68). A purer monoethanolamide may be produced using an excess of monoethanolamine followed by washing (69, 70).

In the US, typical commercial names of sulphated coconut monoethanolamide were Alframine, Xynomine, Alrosene, Hytergen, Sulframine, Miranol, Emcol, and others.

The products are relatively unstable in aqueous paste form and with time hydrolyse to the extent of being unusable. However, the dry product, for example, Waschmittel E used in Germany during World War II (71), is apparently indefinitely stable.

Fatty acids other than coconut were used, for example the sulphated capric/caprilic monoethanolamide (72), and were reported to be excellent foaming agents.

Equally, amines other than monoethanolamine were used, for example 1-amino-2,3-dihydroxypropane, diethanolamine, and trimethylol-amino methane (73). The sulphated fatty acyl derivative of 4-amino-butanol-2 was used in Germany under the trade name Igepon B.

Humectol CX, the sulphated diisobutylamide of oleic acid, was largely used in Germany during World War II as a wetting agent in the bleaching process and as a levelling agent for dyestuffs. The sulphated oleic monoethylanilide (74) was used as an emulsifying agent under the trade name Dismulgan IV.

Sulphated amides have no significant use today.

6.6   SULPHATED OLEFINS

Olefins of hydrocarbon chain corresponding to fatty alcohols were found to give secondary sulphate products with surface active properties like the primary alcohol sulphates (75, 76, 77). Despite the good wetting and detergent properties and the comparatively high stability in alkaline solutions (even at the boiling point and in the presence of free caustic soda), the sulphation of the olefins proved difficult compared to the fatty alcohols. This is probably the reason why these products never attained a significant market penetration, despite some interest shown in their synthesis towards the mid/late 1930s.

Shell commercialised sulphated olefins under the trade name "Teepol" manufactured at the Saint Gobain (France) factory but never produced it in the US.

6.7   SULPHATED FATTY ALCOHOLS (ALKYL SULPHATES)

The sulphuric acid esters of the normal primary aliphatic alcohols have been known since the early days of organic chemistry. Sodium cetyl sulphate was first prepared in 1836 (78).

The commercial introduction of fatty alcohol sulphates around 1930 was prompted by

- The vastly superior water hardness resistance which revolutionised household detergent formulations and washing practices
- The development of inexpensive processes to produce industrial quantities of fatty alcohols

The evolution of the manufacturing technology of fatty alcohols is given in Section 6.9.

6.8   SULPHATED ALKYL, ARYL, AND ALKYLARYL ETHERS

The ethylene glycol ethers of alkyl, aryl, and alkylaryl moieties are discussed in the corresponding section on nonionic surface active agents.

The most practical method to produce ethylene or propylene glycol ethers of fatty alcohols, alkylphenols and naphthols is the direct addition of "n" moles of ethylene oxide or of propylene oxide to one mole of the hydroxyl compounds. The nonionic surface active agents thereby produced were used as such or reacted with sulphuric acid (and, later on, phosphoric acid) to give valuable surface active agents (79, 80, 81). The initial development of this surfactants class was pioneered by I.G. Farben, and the products were commercialised under the trade names of Igepal, Alipal, and Leonil.

Sulphates of alkylated phenols and naphthols (82) were quite common for many years since their market introduction, driven by the lower cost and broader availability of hydrophobic moiety than the fatty alcohols. Concerns on the human and environmental safety of the products and of their metabolites in the last 20 years have progressively reduced their use to industrial processes and formulations in which their replacement could be problematic because of the time and costs involved and, in some instances, the absence of suitable functional replacements.

The manufacturing process and equipment are essentially the same as those used for the alkyl sulphates.

These days, a large volume of the low ethoxylate ("n" 2–3) C12–C14 fatty alcohols are used in personal care (hair and body shampoos), domestic hand dishwashing liquid detergents, and some laundry washing powders and in some industrial applications.

6.9 FATTY ALCOHOLS SULPHATION PROCESS AND RAW MATERIALS

Saturated fatty alcohols react at the terminal hydroxyl group to yield fatty alcohol sulphates and water. There are side reactions leading to olefins, dialkyl sulphate esters, and dialkyl ethers, but these are marginal, and the sulphation reaction can be regarded as quantitative.

The situation is different for unsaturated fatty alcohols that have two possible reaction loci with several reaction possibilities. Riess (83) was the first to investigate systematically the effect of sulphonation conditions. In the case of oleyl alcohol (Delta 9, 10-octadecenol-1), there are four reactions possible:

- Esterification of the terminal hydroxyl, without attacking the double bond
- Sulphuric acid attack at the double bond, without reaction at the terminal hydroxyl
- Double condensation and addition at the terminal hydroxyl and the double bond, leading to octadecan-9,18-diol-disulphate ester
- Further sulphonation reactions at the 9,10 double bond with the ultimate formation of a sulphate/sulphonate triester

The preferred sulphation agent was 98–100% sulphuric acid (84, 85), although oleum of various SO_3 strengths (86, 87, 88) and chlorosulphonic acid also were considered (89).

Instead of sulphuric acid or oleum or chlorosulphonic acid, and sulphamic acid can be used as the sulphating agent. This is discussed in detail in Section 6.16.2. The reaction with sulphamic acid yields ammonium sulphate as a by-product, which is not always acceptable, for example in personal care formulations or emulsion polymerisation recipes.

The first installations for the sulphation process were of the batch type, and reactors of 3000 litres and neutralisers of 5000 litres were installed already in the late 1920s by the company Karl Fisher.

The first patents covering continuous sulphation were assigned to Böhme Fettchemie (90). Rudolf & Co. (91) used tall cylinders with internal wall cooling coils and central helix agitation across the entire height of the reactor. The reactor was filled with a sulphation medium. The fatty alcohol injected at the bottom was finely dispersed and raised by the helixes through the sulphation medium. The sulphated esters were recovered at the top.

An innovative concept, developed by Henkel & Cie, involved the fine dispersion of the reaction mass with a cylinder rotating at 2000–6000 rpm (92).

6.10   ALKYLARYL SULPHONATES

The development of alkylaryl sulphonates resulted from almost parallel work on the alkylation and sulphonation of naphthalene and benzene. There is no clear dividing line between the two, and each advance in one class bore consequences in the other. Thus, the segmentation we propose is somewhat arbitrary, and there may be repetitions or overlapping in the text, but we have decided to retain a separation in view of the diverging paths of the two categories in terms of properties, markets, and applications.

6.10.1   ALKYL NAPHTHALENE SULPHONATES

If we consider the commercial aspects, it can be said that propylated naphthalene sulphonates are the oldest synthetic detergents after sulphonated oils. The interest in the products arose in Germany during World War I to develop substitutes for soap from non-fatty raw materials. However, as discussed hereinafter, the laboratory development of alkyl benzene sulphonates by Krafft and Adams largely predates that of the alkyl naphthalene sulphonates.

Günther and Hetzer must be credited for having laid the foundations of the alkyl-naphthalenes sulphonates chemistry (93). Polynuclear aromatic hydrocarbons were made to react with lower alcohols (especially secondary alcohols, e.g., isopropanol) in a water-stripping medium (e.g., sulphuric acid) in a single-pot, three-stage sequential process. First, the lower alcohol reacts with the sulphuric acid. The ester formed alkylates the aromatic reagent, and the alkylaryl intermediate is eventually sulphonated, yielding a mixture of mono-, di-, and tri-homologues. As can be expected, the di- and tri-propyl naphthalenes are more surface active than the mono. However, the biggest increase in surface activity occurs as the proportion of the di-homologue increases, whereas the tri-homologue has a much smaller effect. Therefore, the commercial products tended to be mixtures in which the di- derivative predominated. The propyl groups are attached through the number 2 carbon, so it would be more correct to speak of isopropyl derivatives. At the time of early manufacturing, the position of substitution on the naphthalene ring was unclear, and it was assumed that the commercial product contained several position isomers.

In improved processes, the one-step reaction was replaced by an initial alkylation (optionally followed by purification), and the alkyl naphthalene was eventually sulphonated. Other than isopropanol, alkylating agents were n-propanol, n-propyl or isopropyl halides, di-isopropyl ether, and propylene. As catalysts, sulphuric acids in various concentrations, $AlCl_3$, $ZnCl_2$, BF_3, and hydrofluoric acid (HF) were proposed (94, 95).

The isopropyl naphthalene sulphonates are poor detergents but are good wetting agents and dispersants and, as such, were extensively used, for example, in janitorial metal cleaning, and crop protection formulations. During World War II, they were used in Germany for the emulsion polymerisation of synthetic rubber.

Among the best-known trade names, there were Alkanol B (I.E. Du Pont de Nemours & Co.), Nekal A (General Dyestuff Corporation and I.G. Farbenindustrie), Aerosol OS (American Cyanamid Corporation), and Naccosol A (National Aniline Division, Allied Chemical and Dye Corp.).

With the years, butylated naphthalene sulphonates (96) became more important commercially and, in part, supplanted the isopropyl products. They were prepared with the same method as for the isopropyl products, using any of the butyl alcohols, although n-butanol was the preferred starting material, and marketed for the same applications. The best-known trade names were Nekal B or BX from the original manufacturer I.G. Farbenindustrie.

A large number of other lower alkyl naphthalenes were proposed, for example mono- and di-amyl and mono- and di-capryl naphthalene sulphonates, mixed dialkyl derivatives, cyclohexyl and methylcyclohexyl derivatives, benzyl derivatives, naphthalene alkylated with various terpenes and terpene alcohols, products made by condensing glycols or alkylene dihalides with naphthalene, and naphthalene-furfural condensates. These products never attained a significant market penetration (if any).

6.10.2 NAPHTHALENE SULPHONATE-FORMALDEHYDE CONDENSATES

Another class of lower alkyl naphthalene sulphonates developed in the 1930s and one that is still largely used today in, for example, SBR polymerisation, pigments dispersion, crop protection formulations, concrete mixtures, and gypsum board slurries includes the condensation products of two or more naphthalene groups joined by methylene groups.

The first patent reference to naphthalene sulphonates formaldehyde condensates is US 1,758,277 of May 13, 1930, to IG Farbenindustrie. The priority date is April 1, 1925, in Germany and March 18, 1926, in the US, which indicate that the supporting development work must go back a couple of years prior to the application.

Naphthalene or lower alkylated naphthalenes may be used in the reaction, which in its simplest manufacturing process consists in heating a mixture of the naphthalene nuclei, formaldehyde, and sulphuric acid or by treating naphthalene sulphonic acids with formaldehyde.

The formaldehyde, naphthalene sulphonate condensates are important for the history of surface active agents for being probably the first example of synthetic oligomeric material recognised and used in industrial processes for their interfacial properties.

6.10.3 ALKYL BENZENE SULPHONATES

Alkyl benzene sulphonates, one of the oldest and most widely used synthetic detergents, are used across all the segments of industry. They were first introduced in the 1930s in the form of branched alkylbenzene sulphonates (97). However, during the 1960s, following environmental concerns, these were replaced with linear alkylbenzene sulphonates (98). Since then, production has increased significantly from about 1 million tons in 1980 to about 3.5 million tons in 2016, making them the most produced anionic surfactants after soaps.

Laboratory processes for the synthesis of alkyl benzene sulphonates were developed already at the end of the 19th century, yet it took about a generation before their surface active properties were understood and recognised.

Krafft (99) synthesised the alkyl benzene by jodation of high-molecular-weight alcohols with hydrogen iodide. These were then converted into alkylbenzene through the Wurtz synthesis with jodobenzene and sodium and finally into alkylbenzene sulphonates by treatment with sulphuric acid.

Adam (100) reacted palmitoyl chloride with benzene using stoichiometric quantities of aluminium chloride catalyst. The resulting benzopalmitophenone was reduced to alkylbenzene with a zinc amalgam and hydrochloric acid and sulphonated with sulphuric acid.

Lindner was the first to appreciate that these high-molecular-weight sulphonates had useful colloidal and cleaning properties (101). He also developed a simpler process than the Krafft or Adam synthesis to produce surface active materials starting from hydrocarbons or phenols and fatty alcohols in the presence of strong condensing agents, like oleum, sulphur trioxide, or chlorosulphonic acid (102, 103, 104). These products were commercialised under the name Melioran and can be considered the precursors of modern alkylbenzene sulphonates.

In 1926, the feasibility of the alkylation of aromatic hydrocarbons with olefins was discovered almost simultaneously in the laboratories of the I.G. Farbenindustrie A.G. (105) and the Chemische Fabrik Pott & Co. (106, 107). The principle was that the olefin double bond is more reactive than the alcohol to alkylate aromatic hydrocarbons in the presence of the additional catalyst, for example, $AlCl_3$, such that the direct alkylation of benzene (as opposed to only naphthalene) is possible.

The pivotal patent for the industrial synthesis of dodecylbenzene from olefins is described in US Pat. 1, 995,827 to Sharples Chemical Co. Although alkylbenzene was used before then, it became commercially relevant only during and after the Second World War, for example as the Oronite Alkane (Oronite Chemical Co., US), the Nacconol NR (National Anyline Division, Allied Chemical and Dye Corp.) and the Santomerse (Monsanto Chemical Co.) that, in 1943, accounted for more than 50% of the tonnage of synthetic detergents produced in the country.

In 1931, Lurie in the US developed a variant of the Krafft process using the Friedel- Crafts synthesis (108). Thus, dodecane was chlorinated, reacted with benzene with the elimination of HCl in the presence of an $AlCl_3$ catalyst and the resulting product sulphonated. It is to be noted that the chlorination of dodecane leads to six possible position isomers and, thus, six different alkylbenzenes. The industrial processes for the manufacture of alkylbenzene from chlorinated hydrocarbons are described in two patents (109, 110).

The preferred catalyst was $AlCl_3$ finely dispersed or solubilised in nitroparaffines (111) or in nitrobenzene. Kerosene or Fischer–Tropsch paraffins were generally used as the starting raw material for alkylation. In Germany, I.G. Farbenindustrie A.G. manufactured at least three products of the higher alkyl aromatic sulphonate series. Igepal NA was produced starting with a hydrogenated C13–14 fraction of Fisher–Tropsch hydrocarbon called Mepasin, particularly rich in linear components. This was chlorinated and condensed with benzene using an $AlCl_3$ catalyst and sulphonated (112). A similar product was made by condensing tetrapropene with benzene and subsequently sulphonating it. The condensation was carried out at 5–10 °C using anhydrous HF as catalyst. The product was made at Hoechst and simply called Ho/1/181. The cyclohexylamine salt of Igepal NA was used as an emulsifying agent under the name Emulphor STL. In another synthesis, the alkylation of benzene with propylene oligomers of boiling range 196–280 °C or 210–282 °C was carried out in 98% sulphuric acid medium which catalysed the reaction (113). Other catalysts used include HF or AlCl$_3$ (114).

Surface active agents of benzene, phenol, and naphthalene were produced by condensation with oleic acid or its esters or amides. Oleic acid reacts readily with benzene giving 9- or 10-phenylstearic acid, which is then sulphonated. The product was claimed to be a lime-resistant wetting agent and detergent. Other aromatics can be used instead of benzene (115, 116, 117).

Branched alkylbenzene sulphonates (BABSs) were first introduced in the early 1930s and saw significant growth from the late 1940s onwards (118). They were prepared by the Friedel–Crafts alkylation of benzene with 'propylene tetramer' followed by sulphonation (119).

The highly branched tail makes BABS difficult to biodegrade (120). It caused the formation of large expanses of stable foam in areas of wastewater discharge such as lakes, rivers, and coastal areas (121), as well as foaming problems encountered in sewage treatment and contamination of drinking water (122). As such BABS was phased out of most detergent products during the 1960s, being replaced with linear alkylbenzene sulphonates (LABSs). It is still important in certain agrochemical and industrial applications, where it proved to be irreplaceable.

The LABSs that replaced the BABSs were first produced from chlorinated paraffins, but this raw material was progressively phased out in favour of olefins. The process and raw material are discussed in Section 6.16.

In the commercial LABS molecule, the sulphonation moiety is mostly in the paraposition of the benzene ring with respect to the alkyl chain, which is attached at any position of the chain, except the terminal carbons. The alkyl carbon chain typically has 10 to 14 carbon atoms, and the linearity of the alkyl chains ranges from 87–98%. While commercial LABSs consist of more than 20 individual components, the ratio of the various homologs and isomers, representing different alkyl chain lengths and aromatic ring positions along the linear alkyl chain, is relatively constant in current industrial products, with the weighted average carbon number of the alkyl chain comprised between 11.7–11.8. LABS is the primary cleaning agent used in many household laundry detergents and cleaners industrial and institutional cleaners and in practically all industrial segments, from crop protection (in the form of calcium salt) to paints, textiles, and many others.

6.10.4 ALKYL DIPHENYL OXIDE (DI)SULPHONATES

A diphenyl oxide precursor is produced commercially by Dow Chemical Company since 1924.

A process for producing diphenyl oxide disulphonates is reported in US 2,081,876 of May 25, 1937, to Du Pont de Nemours. The process involves sulphonating and condensing, in any order, diphenyl oxide with unsubstituted hydroaromatic and aliphatic radicals containing more than two carbon atoms. The surface active properties of the products were recognised, although most of the citations are for low-molecular-weight alkyl chains.

Two years later, Dow was granted a patent (123) covering a process to produce diphenyl oxide disulphonates via the reaction of an alkyl halide, an alcohol or an olefin with diphenyl oxide in the presence of a Friedel-Craft catalyst. The product's applications indicated were as dielectric agents in transformers and plasticisers. Also, this patent puts emphasis in low-molecular-weight alkyl chains.

Higher chain-length hydrocarbon radicals are mentioned in US 2,242,260 of May 20, 1941, to Lubrizol Corporation, which mentions the possibility of using them in formulations that provide enhanced lubricity.

Alkyl diphenyl oxides disulphonates have high hard-water and ionic strength tolerance, are stable in acids and alkalis, and have become established in household detergents and industrial applications like hard surface cleaners (low streaking-filming).

6.11   ALKYL (PARAFFIN) SULPHONATES

The preparation of low-molecular-weight alkylsulphonates was described by Strecker (124) and Haemilian (125) and involved the reaction of a lower alkyl halogenide with sodium sulphite. The method is, however, unsuitable for higher higher-molecular-weight halogenides as alkanes and alcohols are preferentially produced (126).

Paraffin sulphonates are described for the first time in a paper by A. Reychler (127). Cetyl sulphohydrate (prepared from commercial $C_{16}H_{33}$ and an NaSH alcoholic solution) was oxidised with a solution of potassium permanganate. The free acid was recovered after treatment with lead acetate and H_2S and further purified by water washing followed again by a lead acetate and H_2S treatment.

As in many other instances, the discovery of this chemical class was not followed by an attempt to understand the performance and effects of its members. In other words, the researchers were satisfied with the characterisation of the synthetic process and the basic properties of the products as opposed to what they could actually do with it.

Other synthetic routes developed in the following years included the following:

- Reaction of sulphate esters with sodium sulphite (128, 129, 130)
- Hydrohalogenation of alkenes with HOCl and subsequent treatment of the 1-chloro-2-hydroxyalkan with sodium sulphite, leading to beta-hydroxyalkane-alpha-sulphonate (131)
- Halogenated ketones (132); alpha-halogenated fatty acids (133); castor-, oleic-, and ricinoleic acids (134); halogenated carboxylic esters (135, 136, 137); and long-chain alkene ketones (138) react with bisulphite to give alkansulphonates
- Reaction of ethylene and sulphite at 115–125 °C at 700–1000 Atm in carbon tetrachloride, giving products of structure $H(CH_2\text{-}CH_2)_n\text{-}SO_3H$ (139)
- Reaction of alkyl-mercaptans and polysulphides followed by oxidation with potassium permanganate (140). Secondary mercaptans were converted to secondary sulphonates with nitric acid (141)

In 1933 in the US, Reed and Horn discovered that paraffins react with SO_2 and chlorine at low temperatures and under radiation of 3000–3660 Å wavelengths to create alkylsulphochlorides that can be hydrolysed with alkalis to alkyl sulphonates (142, 143). Further work from I.G. Farbenindustrie A.G. solved many of the outstanding synthesis and process problems and was extensively patented (144, 145, 146, 147, 148, 149, 150, 151).

The process from I.G. Farbenindustrie A.G. was accepted by the industry, and during the Second World War, alkyl sulphonates covered a significant proportion of the German demand for detergent raw materials. Leuna Werke alone is believed to have produced more than 50 ktons/year of 100% active alkyl sulphonates. The total production in Germany must have been in the region of 70–100 ktons or higher, and Henkel, for example, produced in their units at Düsseldorf and Genthin some 250 ktons/year of detergent compositions based on alkyl sulphonates.

The industrial state of alkane sulphonates in Germany in the late thirties and during World War II has been described by Hoyt (152). A fraction of saturated Fischer–Tropsch hydrocarbons averaging 15 carbon atoms was converted to the crude sulphonyl halide with sulphur dioxide and chlorine. Two types of crude sulphonyl chloride were produced: Mersol D (50% monosulphonyl chloride, 30% disulphonyl chloride) and Mersol H (50% monosulphonyl chloride, 50% of the unreacted hydrocarbon). The crude Mersols were supplied to detergent manufacturers who hydrolysed them with caustic soda and separated the unreacted hydrocarbons to whatever degree was most practical. The Mersols were also used as intermediates for reaction with sarcosine, taurine, and others to make other types of surface active agents, but the volumes for those applications were small. A saponified product essentially free of hydrocarbons was commercialised by I.G. Farbenindustrie A.G. under the trade name of Mersolat H.

The sulphoxidation synthetic route to alkyl sulphonates was discovered by Platz and Schimmelschmidt at I.G. Farbenindustrie Werke Hoechst (153) and developed in conjunction with Leuna between 1941–1945 (154).

Alkyl sulphonates produced through the sulphoxidation process never reached the widespread use of sulphochlorination products like Mersolat (Werk Leuna, Bayer Leverkusen), Witolat (Imhausen-Witten), Mainolat (Bayrischen Rohstoffwerken), and MP 189 (Du Pont USA).

During World War II, alkane sulphonic acids were produced on the pilot plant scale in Germany directly from paraffin hydrocarbons, sulphur dioxide, and oxygen. This eliminates the necessity for using chlorine to form the intermediate sulphonyl halide. In variations of the process, ozone and actinic light or organic peracids were used as activating agents (155, 156).

6.12   OLEFIN SULPHONATES

The sulphonation of olefins or fatty acids is an extension of the sulphation process, conducted with more energetic sulphonating agents.

Alpha-olefin sulphonates are obtained by sulphonation of alpha-olefins with gaseous SO_3 followed by isomerisation and hydrolysis of the reaction product (157). According to Brit. Pat. 1,174,857 (to Colgate-Palmolive), olefin sulphonates are prepared by (1) reacting diluted gaseous sulphur trioxide and an olefin to produce an acid mix; (2) neutralising the acid mix with an aqueous alkaline material while maintaining the temperature below 65 °C, the amount of alkaline material and water being such that the resulting blend has a pH of at least 12 and a viscosity of at least 5000 centipoises; and (3) passing the viscous alkaline blend continuously into a zone maintained under super-atmospheric pressure in which the blend makes contact with a heated solid heat exchange surface maintained at a temperature of at least 175 °C so as to raise the temperature of the blend to at least 165 °C within a period of less than 5 minutes. The alkaline material is generally used in a quantity of 110–140% of the stoichiometric amount required. Suitable olefins contain 8–30 C atoms. The acid mix may also be prepared by reacting the SO_3/olefin product with concentrated H_2SO_4.

Hydrolysis is necessary because after sulphonation, a mixture of alkene sulphonic acids and sultones is obtained. Sultones isomerise and hydrolise at high temperatures to hydroxyalkane sulphonate. The final product typically contains about 70% of an olefin sulphonate:

$$CH_3(CH_2)_nCH = CH(CH_2)_mSO_3^-M^+$$

and about 30% of an hydroxyalkane sulphonate

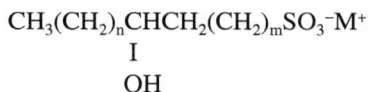

$$CH_3(CH_2)_nCHCH_2(CH_2)_mSO_3^-M^+$$
$$|$$
$$OH$$

Direct sulphonation of an aliphatic hydrocarbon with sulphuric acid is difficult. Saturated fatty acids, fatty acid esters or anhydrides, and fatty acids sodium salts were, however, readily and directly sulphonated to give α-sulphonic acids using, for example, sulphur trioxide in liquid sulphur dioxide (158). Saturated aliphatic ketones were sulphonated with oleum and acetic anhydride or by first halogenating and then replacing the halogen atom with sodium sulphite (159).

Strong sulphonating agents like sulphur trioxide or chlorosulphonic acids were found to react with olefins in the surface active range (e.g., 1-hexadecene) or with fatty alcohols to give true sulphonic acids, stable in hot mineral acids solutions, where esters would typically hydrolise (160).

6.13   METHYL ESTERS SULPHONATES

In the mid-1950s/early 1960s, the US Department of Agriculture attempted to develop tallow methyl ester sulphonates (MESs) as a means for adding value to tallow fats. Stirton et al. produced several publications, and patents were filed from detergent manufacturers (161).

A schematic picture of an MES is shown below

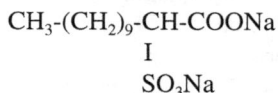

$$CH_3-(CH_2)_9-CH-COONa$$
$$|$$
$$SO_3Na$$

The technology available at the time and for many years thereafter did not allow producing surface active agents with sufficient consistency of composition, appearance, and organoleptic properties to be acceptable to formulate household detergents or for industrial applications. Residual impurities caused sensitisation problems with consumers. It is only in recent years that improvements in the manufacturing process have overcome these drawbacks (162, 163, 164).

The modern manufacture of MESs consists of three stages:

- A methyl ester/SO_3 contacting stage in which 1.2 moles of SO_3 are reacted with one mole of methyl ester (ME). The excess SO_3 is necessary to achieve full conversion of the ME. This stage is usually carried out continuously in a falling film reactor.
- An ageing stage in which the intermediate species formed in stage one rearrange and the conversion of ME to MES goes to completion. The residence time depends on the temperature, the mole ratio SO_3/ME, the reactor characteristics, and the degree of conversion targeted.
- A neutralisation stage usually is carried out continuously in a loop reactor, avoiding extremes of pH that would otherwise back-hydrolise the MES to the di-salt. Neutralisation is usually carried out in a 60% active matter paste, which contains MES and the di-salt $RCH(CO_2Na)SO_3Na$ in proportions of about 80/20. Sodium methyl sulphate is also present, approximately equimolar with the di-salt.

Depending on the formulations in which MES is to be used, this level of purity may or may not be acceptable. If higher purity MES is wanted, there are two additional steps needed:

- A bleaching step, which is a must for products used in laundry detergents
- A re-esterification step to convert the di-salt precursor into an MES precursor. This consists of treating the acidic mixture with methanol before neutralisation and reduces the di-salt content in the neutralised final product to below 10% based on a 100% active matter (165).

A drawback of the MES is the hydrolytic stability. Although a combination of steric and electronic effects of the α-sulpho group slows down the rate of hydrolysis compared to the non-sulphonated esters, MES are not suitable for formulation by the spray drying of an aqueous slurry. They can, however, be used as the dried solid for the non-tower production of detergent powder.

The interest in MES has been fuelled in the recent past because of the following:

- Their sustainability
- The cost advantage over linear alkylbenzene sulphonate
- The possibility of using ME streams unsuitable for biodiesel

It is our opinion that only sustainability is a real differentiator and advantage. Biodiesels based on fatty acids MEs are expensive and likely to be supplanted by bioethanol once enzymes converting directly cellulose to ethanol become commercially available (and these are in an advanced stage of development). The cost advantage over LBASs is equally likely to vanish as there is any reason to believe that with time the price of renewable raw material will align with that of petrochemical feedstock.

At the present, MESs are produced in Japan by Lion (40 ktons/year capacity), the US by Stepan (50 ktons/year capacity), and the US by Huish (80 ktons/year capacity). Stepan has developed a high-purity grade which has been approved for and is being used in baby shampoo formulations. Lonkey Industrial Company Ltd. in the People's Republic of China is reported to have completed an MES plant in 2008.

6.14   ISETHIONATES

Two general classes of ester-linked surface active agents in the alkane sulphonate class were developed in the early 1940s. In the first class, the hydrophobic radical is a fatty acid, which is esterified with the hydroxyl group of a low-molecular-weight aliphatic hydroxy sulphonate. In the second (and less common) class, the hydrophobic radical is a fatty alcohol, which is esterified with a low-molecular-weight aliphatic carboxy sulphonate.

The oldest and most popular products are of a general formula:

$$RCOOC_2H_4SO_3Na$$

They were made by direct or indirect esterification of a fatty acid with sodium isethionate $HOC_2H_4SO_3Na$ (in turn obtained by the direct ethoxylation of sodium bisulphite in aqueous solution at 70–80 °C), although other processes were proposed:

- Reaction of a fatty acid with a carbyl sulphate (the anhydride of the sulphuric ester of isethionic acid)
- Reaction of the sodium salt of a fatty acid with chloroethane sulphonic salts (166)
- Reaction of the chloroethyl ester of a fatty acid with sodium sulphite (167)
- Transesterification of a low-molecular-weight isethionic ester, for example, acetylisethionic acid and a higher fatty acid (168)

The synthetic route via fatty acids acyl chlorides has been abandoned.

I.G. Farbenindustrie A.G. took several patents on the manufacturing process (169, 170, 171, 172).

Fatty isethionates were commercialised in various physical forms and strengths by I.G. Farbenindustrie A.G. and Aniline and Film Corp. as Igepon A, AP, and AP Extra.

As stated earlier, the ester-linked alkane sulphonated produced by reacting first a fatty alcohol with chloroacetic acid or chloroacetyl chloride and then with an alkali metal sulphite were less common and never achieved substantial market penetration. Nacconol LAL, sold by the National Aniline Division of Allied Chemical and Dye Corp., enjoyed some popularity as an ingredient in toothpaste, shampoo, and personal care formulations because of its cleansing and foaming properties and its colourless, clean appearance. Its high manufacturing costs, however, prevented more widespread use.

Today, large volumes of isethionates are used for the formulation of synthetic and combo bar soaps, where they can effectively replace fatty alcohol sulphates because of the reduced water solubility. This prevents the rapid consumption of the bar and an aesthetically objectionable softening and swelling (in German, *versumpfung*). The personal care industry also is leaning more and more towards the isethionate, in part because of advances in formulation technologies (which mitigate the poor water solubility), the development of more soluble molecules (methyl isethionates), and the "sulphate free" drive.

At the time of compiling this book, there were several suppliers of isethionates but only one for methyl isethionate (Innospec), which was sold under the trade name of Iselux.

6.15   DICARBOXYLIC (SULPHOSUCCINATES) AND TRICARBOXYLIC SULPHONATED ESTERS

Hydrophobic esters having a double bond adjacent to an activating carboxylic ester group can easily add $NaHSO_3$, and among these addition products, there are the sulphosuccinates, some of the best-known and more widely used wetting agents.

Sulphosuccinates were (and are) prepared by reacting a fatty alcohol (optionally alkoxylated) with maleic acid. Depending on the molar ratio mono- or diesters are made. These are treated with concentrated aqueous solutions of bisulphite, which adds to the double bond and forms mono- or di-alkyl sulphosuccinates (173, 174, 175).

The sulphosuccinates' chemistry allows for many variations in terms of the degree of esterification and the type of hydrophobe. Other than the simple alcohols of different chain lengths, saturation/unsaturation/substitution at the alkyl chain of more complex alcohols may be used to provide the hydrophobic moieties. Typical examples are alkoxylated fatty alcohols, hydroxyethyl amides of fatty acids (176), and ricinoleic acid esters (177). Over the years, this has favoured the development of many products with different and specific characteristics for different applications, for example, mild cleansers in hair and body shampoos, wetting agents in crop protection formulations, pigment dispersions, and oil-slick dispersants. Furthermore, sulphosuccinates are colourless, have low mammalian toxicity, can be supplied in aqueous and non-aqueous solutions of different strengths, and can be considered essentially renewable.

For many years, the sulphosuccinates market was dominated by American Cyanamid Co., and the trade names Aerosol and Deceresol have been reference industry standards. Even today, the American Cyanamid offspring Cytec enjoys a position of market leadership and offers a much more complete range of products than other manufacturers that focus on the products demanded in larger volumes or that are easier to manufacture.

Esters of sulphotricarboxylic acids were also developed around the end of 1930 (178, 179). A commercial wetting agent of this family was Nekal NS (General Dyestuff Corp.), the trihexyl ester of sulphotricarballylic acid and produced by dehydrating trihexyl citrate to form trihexyl aconitate and then adding sodium bisulphite across the double bond.

In the early days of synthetic detergents, rosin, crude coal tar, and high-boiling coal tar (all containing condensed aromatic rings in various proportions) were sulphonated to yield products with mild surface active properties (180, 181, 182, 183).

6.16   SULPHONATION PROCESS AND RAW MATERIALS

6.16.1   ALKYLATION OF AN AROMATIC MOIETY

Probably the oldest and best-known procedure for alkylating an aromatic moiety is the classical Friedel-Crafts synthesis. In this reaction, alkyl halide RX reacts with an aromatic hydrocarbon ArH in the presence of anhydrous aluminium chloride (184).

$$RX + ArH \rightarrow RAr + HX$$

In place of alkyl halides alcohols, olefins, ethers, dialkyl sulphates, alkyl sulphuric esters, alkyl sulphonates, phosphates and other derivatives can be used, all of which function in this reaction as the equivalent of an alkyl halide (185). Catalysts other than aluminium chloride have been used: gaseous hydrochloric acid (HCl), sulphuric acid, boron fluoride, HF, phosphoric acid, phosphoric anhydride, and the halides of zinc (186).

Based on the experimental evidence available at the time, the mechanism of the alkylation reaction was described theoretically with a carbonium ion mechanism (187, 188). Carbonium ions may form from halides

$$RX \rightarrow R^+ + X^-$$

Or, in the case of olefins,

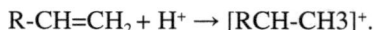

$$R\text{-}CH\text{=}CH_2 + H^+ \rightarrow [RCH\text{-}CH3]^+.$$

The carbonium ion is an electrophilic substance and attacks the aromatic moiety by displacing a proton in the position where the electron density is greatest (189). This is, of course, irrelevant in the case of non-substituted benzene.

In Germany, I.G. Farbenindustrie A.G. manufactured at least three products of the higher alkyl aromatic sulphonate series. Igepal NA was produced starting with a hydrogenated C13–14 fraction of Fisher–Tropsch hydrocarbon called Mepasin. This was chlorinated, condensed with benzene using an $AlCl_3$ catalyst and sulphonated (190). A similar product was made by condensing tetrapropene with benzene and subsequently sulphonating it. The condensation was carried out at 5–10 °C using anhydrous HF as a catalyst. The product was made at Hechst and was simply called Ho/1/181. The cyclohexylamine salt of Igepal NA was used as an emulsifying agent under the name Emulphor STL.

Surface active agents of benzene, phenol, and naphthalene were produced by condensation with oleic acid, or its esters or amides. Oleic acid reacts readily with benzene giving 9- or 10-phenylstearic acid, which is then sulphonated. The product was claimed to be a lime-resistant wetting agent and detergent. Other aromatics can be used instead of benzene (191, 192, 193).

The alkylating agent mostly used to produce BABSs is the propylene tetramer obtained by continuous polymerisation at 1.5 MPa and 200 °C with orthophosphoric acid or zeolite catalysts. The reaction produces a mixture of olefins, from which the dodecane cut is isolated by distillation.

The alkylating agents needed to produce LABS can be obtained in several ways (194).

The Friedel-Crafts alkylation process involves chlorinating n-paraffins into monochloroparaffins followed by alkylating benzene using aluminium chloride catalyst. This method is one of the oldest commercial routes to linear alkyl benzene (LAB) and has been almost completely abandoned.

The HF/n-paraffins process involves the dehydrogenation of n-paraffins to olefins and the subsequent reaction with benzene using hydrogen fluoride as catalyst. This process accounts for most of the installed LAB production in the world. It includes a PACOL stage, where n-paraffins are converted to mono-olefins (typically internal mono-olefins); a DEFINE unit whose primary function is to convert residual diolefins to mono-olefins; a PEP unit, which is essentially an aromatic removal unit, introduced before the alkylation step to improve LAB yield and quality, an alkylation step where mono-olefins, both internal and alpha-olefins, are reacted with benzene to produce LAB in the presence of an HF catalyst.

The newer DETAL process (introduced in 1995) has several of the stages of the HF/n-paraffins process, but it is principally different in the benzene alkylation step, during which a solid-state catalyst is employed. The process eliminates catalyst neutralisation and HF disposal. Consequently, most LABS plants built since then have utilised this process. There is a further improvement of the DETAL process that includes a step in which any higher alkylated benzenes are contacted with additional benzene over a transalkylation catalyst.

Each process produces LAB with distinct features. Important product characteristics include the bromine index, the sulphonatability, the amount of 2-phenyl isomers (2-phenylalkane), the tetralin content, the amount of non-alkylbenzene components, and the linearity of the product.

The production of n-paraffins often occurs as part of an integrated LAB plant where the producers start with kerosene as raw material. The process for producing normal paraffin includes a kerosene prefractionation unit, a hydrotreating unit and a unit that separates n-paraffins from isoparaffins with 98% recovery of 99% pure n-paraffins (UOP Molex unit). The ExxonMobil Chemical technology includes a recovery process and can produce LAB grade n-paraffins from most medium- to low-sulphur kerosene without the use of a hydrotreater stage upstream. A desulphurisation process is needed to reduce the sulphur content of some n-paraffins.

6.16.2 Sulphation and Sulphonating Agents

The sulphonating agents described here are suitable for any sulphonation and sulphation reaction, with the exception of sulphamic acid, which can be used only for sulphation processes.

SO$_3$ is an aggressive electrophilic reagent that rapidly reacts with any organic compound containing an electron donor group. Sulphonation is a difficult reaction to perform on an industrial scale because the reaction is rapid and highly exothermic, releasing approximately 380 kJ/kg SO$_3$ reacted. Most organic compounds form a black char on contact with pure SO$_3$ due to the rapid reaction and heat evolution. Additionally, the reactants increase in viscosity between 15 and 300 times as they are converted from the organic feedstock to the sulphonated derivatives. This large increase in viscosity makes heat removal difficult. Effectively cooling the reaction mass is essential because high temperatures promote side reactions that produce undesirable by-products. Also, precise control of the molar ratio of SO$_3$ to organic is essential because any excess SO$_3$, due to its reactive nature, contributes to side reactions and by-product formation. Therefore, commercial-scale sulphonation reactions require special equipment and instrumentation that allows tight control of the mole ratio of SO$_3$ to an organic substrate and rapid removal of the heat of the reaction.

Historically, the problem of SO$_3$ reactivity has been solved by diluting and/or complexing the SO$_3$ to moderate the rate of reaction. Commercially, diluting or complexing agents include ammonia (sulphamic acid), hydrochloric acid (chlorosulphonic acid), water or sulphuric acid (sulphuric acid or oleum), and dry air (air/SO$_3$ film sulphonation).

To achieve the desired product quality, controlling the ratio of SO$_3$ to organic raw material is vital, as is the equipment design to ensure rapid heat removal.

The choice of the appropriate sulphonating agent depends on the type of product and the required product quality. Sulphamic acid cannot be used for sulphonation reactions, Sodium alkyl benzene sulphonate made with oleum contains a minimum of 8% of sodium sulphate. Air/SO$_3$ can sulphonate a wide variety of feedstocks and produces excellent-quality products from all of them.

The air/SO$_3$ process is a large-scale continuous process best suited to 24-hours-a-day, 7-days-a-week manufacture of tons per hour of product. It has the lowest cost per ton of SO$_3$ reacted. The chlorosulphonic acid and oleum processes can be run as either batch or continuous processes. For large-scale commodity production, the air/SO$_3$ process clearly has an advantage. However, for small-scale production of a high-value specialty product, this advantage may be outweighed by other considerations, such as initial equipment cost and the necessity for continuous operation.

The air/SO$_3$ process has the highest equipment costs.

The chlorosulphonic acid and oleum sulphonation processes produce large by-product streams of either hydrochloric acid or sulphuric acid. These by-products must be recovered and sold (which is obviously a distraction from the mainstream production of surfactants) or disposed of as waste.

Over the years, the air/SO$_3$ process has overtaken the oleum or chlorosulphonic acid processes as the predominant choice. This is because of the following:

- Waste disposal cost for the spent sulphuric acid from the oleum process
- The dislike of many processors to hazardous materials such as oleum or chlorosulphonic acid
- The air/SO$_3$ process is capable of making a broad range of very high-quality products
- The move towards compact detergent products which require reduced levels of or no sodium sulphate

6.17 PHOSPHATE ESTERS

Phosphate esters are a versatile range of specialised anionic surfactants displaying many properties which are not possessed by other anionics. Other than the ester bond, this is due to the huge raw material base (alcohols, alkyl phenols of different hydrocarbon chain lengths, as such, and with various degrees of, ethoxylation) and the numerous process routes. Because of this, their consumption has constantly increased, and new applications are constantly emerging.

The chemistry of phosphate esters can be traced back to the second half of the 19th century when in 1883, Weger reported the synthesis of trimethyl phosphate, which, however, apparently had been achieved even earlier by Lossen, as reported by G.A. Petroianu (195).

Early references to surface active phosphate esters can be found, for example, in D.R.P. 575.660 (introduced in 1926) which describes wetting and cleaning agents prepared by the reaction of vegetal or animal oils and fats containing double bonds or hydroxy groups, or their fatty acids, with P_2O_5 or acetylphosphoric acid. The products could be used in the textile industry in place of Turkey Red Oil.

D.R.P. 619.019 (introduced in 1929) describes the preparation of acid phosphoric acid esters by reacting aliphatic alcohols containing more than eight carbon atoms with acetylphosphoric acids or their components.

Mixed phosphates and sulphonated esters are described, for example, in D.R.P. 664.514 (1938) while the phosphate and acetate esters described in French Pat. 642.392 are claimed to have properties similar to Turkey Red Oil.

Improvements in the production process and the quality of the resulting products are described in USA (1934).

From the initial products developed for the textile industries, different phosphate esters molecules and synthetic processes using a variety of raw materials have been proposed over the years, and these have found an increasing number of applications in household, personal care, crop protection, industrial and institutional cleaning, and many other industries. It is an almost impossible task to document each stage of this continuous evolution, and what is reported here is a picture of the spectrum of applications in which phosphate esters are used today.

Today, surface active phosphate esters are used as follows:

- Emulsifiers in general (196). Emulsifiers in emulsion polymerisation for example as described in US 3,963,688 A (1974) that claims the mono- or diester of phosphoric acid and a straight chain alkanol of 8 to 10 carbon atoms is used as a surfactant in the emulsion polymerisation of olefinic compounds, especially vinyl compounds, to obtain polymers having improved stability to light and heath.
- Emulsifiers in personal care preparations: phosphate monoesters have gained positions in the personal care industry as their emulsification properties combine for exceptional skin mildness (197, 198, 199).
- Dispersants: phosphate esters and in particular tristyryl phenol (200) and its ethoxylates are the primary dispersants and emulsifiers for formulating suspension concentrates and water-soluble granules. They are reported to increase the ζ-potential of TiO_2 particles (201).

- Antifoams (202, 203, 204).
- Antistatic agents for synthetic fibres, first reported for rayon fibres in 1951 to 1953 (205, 206).
- Plasticisers: other than the plasticising effect, phosphate esters have fire-retardant properties which are especially needed in demanding applications to improve the behaviour of plasticised polymers. Triaryl phosphates (e.g., cresyl diphenyl phosphate) and alkyl diaryl phosphates (e.g., 2-ethylhexyl diphenyl phosphate) are the two main categories of this kind of plasticiser. While aromatic phosphate esters are typically used in polyvinyl chloride (PVC), theromplastic polyurethanes, thermosets and rubbers, aliphatic phosphates are mainly used in polyurethane foams. In PVC, the phosphate esters offer good gelling behaviour and low temperature performance (207).
- Lubricants: these are mainly aromatic esters and have the additional advantage of providing fire retardancy and some degree of corrosion inhibition.
- Hydrotropes: the hydrotropic properties of phosphate esters allow to produce clear, stable liquid formulations of surface active agents with highly concentrated alkali or other inorganic builders (208).

Against many useful properties of phosphate esters stand nowadays emotional and technical issues. First, "phosphate" sounds bad in consumer marketing terms, and there isn't good biodegradability data to their defence. Second, the mining of phosphorus is quite "unecological". Massive amounts of gypsum are created, contaminated with As, Cd, U, and Ra. Last, the phosphation of ethoxylates with P_2O_5 generates a lot of dioxane that needs stripping.

The reaction of phosphation, sometimes referred to as "phosphorylation", proceeds via the addition of a hydroxyl compound to specific phosphating agents.

The following phosphating agents are the most commonly used today (Figure 6.1):

- Phosphoryl chloride (the $POCl_3$ process)
- Phosphorous pentoxide (the P_4O_{10} process)
- Orthophosphoric acid (H_3PO_4)
- Polyphosphoric acid (the "TPA (tetraphosphoric acid) process")

There is considerable variation in the quantitative composition of reaction products from each process. These include varying levels of mono-, di-, and triester phosphates; unphosphated material; and free phosphoric acid. Only the mono- and diesters have useful surface active properties (Table 6.1).

As a general rule, monoesters are excellent hydrotropes and have high electrolyte tolerance and fair detergency and wetting and are high-foaming and good emulsifiers but poor dispersants. They are exceptionally mild to the skin. Diesters are poor hydrotropes, have poor electrolyte tolerance, are good detergent and excellent wetting agents, are low-foaming, and are good emulsifiers and excellent dispersants.

The typical properties of phosphate esters, for example, foaming/defoaming, emulsification, dispersion wetting, detergency, lubricity, hydrotropicity, and electrolyte tolerance, can be adjusted by choosing the hydrophobe (fatty alcohol type and alkyl phenol, hydrocarbon chain length, degree of ethoxylation).

$$\begin{array}{c} OR \\ / \\ O=P-OH \\ \backslash \\ OH \end{array}$$

Monoester

R - OH + Phosphating agent   → $\begin{array}{c} OR \\ / \\ O=P-OR \\ \backslash \\ OH \end{array}$

Diester

$$\begin{array}{c} OR \\ / \\ O=P-OR \\ \backslash \\ OR \end{array}$$

Triester

Free R - OH
Free H_3PO_4

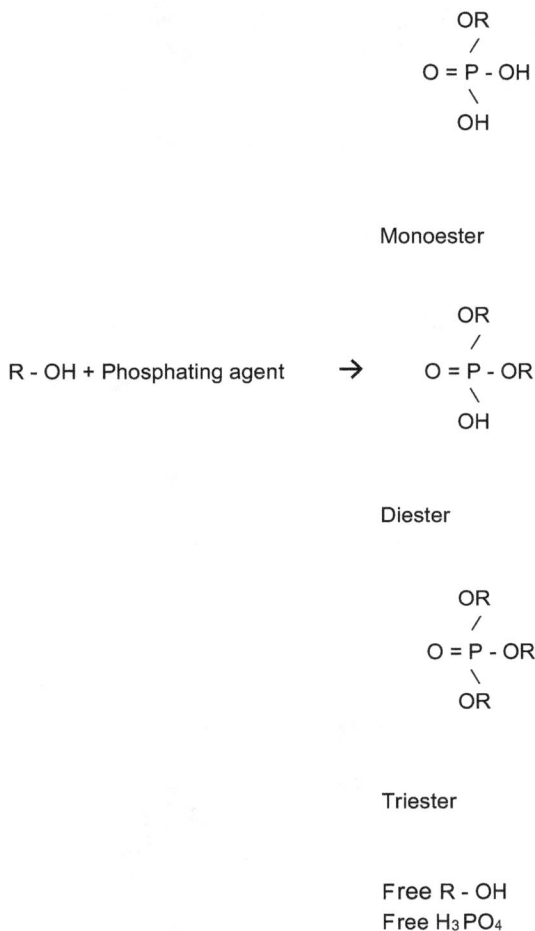

FIGURE 6.1   The phosphation reaction.

Source: Author contribution.

TABLE 6.1
Mono-, Di-, Triester Distribution by Process Type

Process	Monoester % by weight	Diester % by weight	Triester % by weight	Free H_3PO_4 % by weight	Free Nonionic % by weight
$POCl_3$	5 to 10	43 to 48	24 to 28	total 15	total 15
P_4O_{10}	40 to 50	38 to 54	negligible	total 5 to 12	total 5 to 12
H_3PO_4	5 to 10	45 to 50	24 to 28	total 15	total 15
TPA	70 to 90	below 2	below 2	2 to 10	5 to 20

This almost infinite variety of possible combinations has intriguing consequences. For example, it has been observed that a C_{13}–C_{15} alcohol 7 ethylene oxide
(EO) produced by the TPA route has similar surface active properties as a branched
C_{13} alcohol 10 EO produced via the P_4O_{10} route, and both perform very much the
same as emulsifiers in emulsion polymerisation (209, 210, 211).

6.18   SURFACTANTS BASED ON AMINO ACIDS AND PROTEIN HYDROLYSATES

The concept of using amino acids as a hydrophilic moiety was first published in the
early 1930s in Ger. Pat. 546,942, which describes the condensation product of aspar-
tic acid with fatty acyl chlorides to produce surface active agents suitable for wetting
out in mercerising baths. Another early patent (212) describes a detergent and wetting agent resistant to lime and acids made by condensing a fatty acid acyl chloride
with N-β-hydroxy ethylglicine.

The exploitation of the acid halides' reactivity was fundamental for the development of the amino acids and protein hydrolysate chemistry.

Fatty acids can be made more reactive, by turning them into the following:

- Anhydrides

$$2RCOOH \rightarrow RCO\text{-}O\text{-}OCR + H_2O$$

- Acyl halides, most commonly acyl chlorides

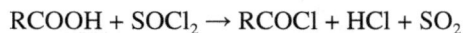

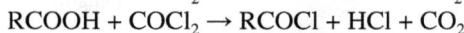

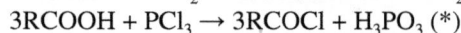

$$RCOOH + SOCl_2 \rightarrow RCOCl + HCl + SO_2$$
$$RCOOH + COCl_2 \rightarrow RCOCl + HCl + CO_2$$
$$3RCOOH + PCl_3 \rightarrow 3RCOCl + H_3PO_3 \; (*)$$

(*) Instead of PCl_3, $OPCl_3$ or PCl_5 can be used.

The use of an acyl chloride instead of the parent fatty acid enables a faster reaction
connection between the fatty acid function with a substrate with labile hydrogens,
such as the following:

- Alcohols (including polyhydric alcohols and sugars) to form esters

$$RCOCl + R'OH \rightarrow RCOOR' + HCl$$

- Amines (primary and secondary) to form amides. This is the Schotten-
  Baumann reaction, described for the first time in 1883 and reported in later
  publications (213).

$$RCOCl + R'NH2 \rightarrow RCONHR' + HCl$$

HCl has to be neutralised as produced with NaOH to avoid yield reduction
by its reaction with the free amine forming the quaternary $R'NH_3^+Cl^-$.

The sulphate-free movement that is gaining momentum in personal care is generating a growing interest in amino acid surfactants, despite their comparatively high cost.

6.18.1 ACYL TAURATES

The condensation of fatty acid chlorides with taurine or N-methyl taurine was developed by I.G. Farbenindustrie A.G. in Germany around the 1930s (214, 215, 216). Taurates were first obtained by the Schotten-Baumann reaction, condensing long-chain carboxylic acid chlorides with aqueous solutions of the sodium salt of N-methyltaurine.

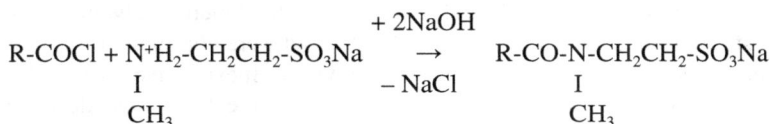

$$
\begin{array}{ccc}
 & + 2NaOH & \\
R\text{-}COCl + N^+H_2\text{-}CH_2CH_2\text{-}SO_3Na & \rightarrow & R\text{-}CO\text{-}N\text{-}CH_2CH_2\text{-}SO_3Na \\
\quad | & - NaCl & | \\
\quad CH_3 & & CH_3
\end{array}
$$

The Schotten-Baumann synthesis of taurates.

This synthesis pathway is complicated and expensive. An advantage, however, is the very low content of residual free fatty acids in the product.

Taurates can also be produced by direct amidation of N-methyltaurine or its sodium salt with the corresponding fatty acid for 10 hours at 220 °C under nitrogen (217, 218).

N-methyl taurine can be produced by reacting methylamine with the following:

- 2-Chloroethane sulphonic acid
- Vinyl sulphonic acid
- Carbyl sulphate

A simpler, more effective process developed in later years involved reacting methylamine with sodium isethionates at 270–290°C under 20–21 MPa. Methyl taurine was used as a 25% aqueous solution directly in the condensation step.

Oleyl taurine was commercialised under the trade name of Igepon T and enjoyed a wide and rapid acceptance in the textile industry, in particular because, as opposed to soap, it does not felt wool during washing. Even in recent years, hundreds of tons were used for textile applications (wetting agents and detergents, dye dispersants).

Taurates are high foaming, even at low pH (down to 4), where all other amino acids surfactants fail. Thy are used as mild, high-foaming surfactants in body cleansing and personal care products (shampoos, liquid soaps and cleansers, face lotions, skin creams, bubble baths, syndet soaps, toothpastes) and in crop protection for spray drying formulations.

Fatty acid chlorides other than oleyl were used in commercial products, for example, coconut (Igepon KT) and myristic/palmitic (Igepon 72 K). Alipon OT was made from oxidised paraffin wax acids. In Eastern and Central Europe (USSR and DDR, respectively), they were formulated in household detergents because of the large manufacturing capacity of the starting raw material at the Leuna Werke Kombinat.

Also, different synthetic processes for alternative taurate structures have been proposed:

- Condensation of fatty acid chlorides with amino methane sulphonates (219)
- Condensation of oleic amide with formaldehyde and N-methyl taurine (220)
- Condensation with β- amino-β'- sulphodiethyl ether (221, 222)
- Condensation of hydroxymethyl lauramide with 2 mercapto ethanesulphonate and many more of dubious commercial and technological interest

6.18.2 Acyl Sarcosinates

The condensation products of acyl chlorides with *N*-methyl glycine (sarcosine) were first reported around 1932 (223, 224, 225) and were commercialised under the Medialan trade name. At the time, sarcosine was an inexpensive chemical, largely available in Germany. It was produced from methylamine, formaldehyde, and hydrogen cyanide and was used as an intermediate for the synthesis of fast cotton dyestuffs. The comparatively low cost of the raw materials was an important contribution to the commercial success of the sarcosinates, which added to their cleansing power, mildness, foaming, water solubility, and lime resistance.

$$CH_3\text{-}(CH2)_{10}\text{-}CO\text{-}N\text{-}CH_2\text{-}COONa$$
$$I$$
$$CH_3$$

Formula of lauroyl sarcosinate

Other than in the textile industry, sarcosinates became popular also in personal care formulations such as hair and body shampoos, shaving foams, and cleansing products, because of their foaming and mildness. Sodium lauroyl sarcosinate was sold as a special ingredient called "Gardol" in Colgate "Dental Cream" toothpaste from the 1950s through the mid-1960s in the US and in the mid-1970s in France and other Western European countries. It is currently used as a preventive dentifrice in Arm & Hammer Baking Soda Toothpaste.

6.18.3 Acyl Glutamates

Acyl glutamates are produced via the Schotten-Baumann reaction of a fatty acid chloride and glutamic acid, an amino acid present in structural proteins (collagen) of human skin (226).

The schematic synthesis of an acyl glutamate is shown below

$$R\text{-}COCl + HOOC\text{-}CH_2CH_2CH\text{-}COOH \rightarrow + HOOC\text{-}CH_2CH_2CH\text{-}COOH + HCl$$
$$I \qquad\qquad\qquad\qquad\qquad I$$
$$NH_2 \qquad\qquad\qquad\qquad\qquad NH\text{-}COR$$

Synthesis of acyl glutamates.

Since the 1970s, acyl glutamates are at the origin of the "mild skin cleansing" concept. The first grade of fatty acyl glutamate was launched by Ajinmoto in 1972 and was used for the first time in a dermatologically cleansing product by Yamanouchi Pharmaceutical Company in Japan.

A sustained interest in acyl glutamates arose in the 1990s for applications in rinse-off products. Today, acyl glutamates are recognised from the scientific point of view as soft, multifunctional surfactants. In Europe, Beiersdorf was among the first European groups to use it in its products on the mass distribution network. Zimmer & Schwartz was the first European surfactants manufacturer to produce acyl glutamates in the Italian plant in Tricerio.

Fatty acid acyl glutamates are present in toothpaste; anti-dandruff, sensitive skin, and baby shampoos; shower creams and gels; and face cleansing foams and gels. Their major drawback is the moderate foaming power which requires foam boosters like taurates or sarcosinates in the formulations. Also, solvents like hexane, acetone, propylene glycol, and isopropyl alcohol may be used in industrial production, which may cause problems for their "green" certification. However, solvent-free synthetic processes have been recently developed (227).

6.18.4 ACYL GLYCINATES

$$RCO-NHCH_2.COONa$$

Acyl glycinate.

Despite of the similar chemical structure, acyl sarcosinates and acyl glycinates have significant differences in the physico-chemical properties in aqueous solutions. This is due to the ability of acyl glycinates to create intermolecular hydrogen bonds between the amide groups, a property that is lost when the amide hydrogen is replaced with an *N*-methyl group in acyl sarcosinates. These attractive interactions between acyl glycinate molecules result – in combination with the absence of the bulky *N*-methyl groups – in a more dense and more stable 'packing' of acyl glycinates in comparison to acyl sarcosinates. For aqueous solutions, this means a higher Krafft temperature of acyl glycinates.

6.18.5 OTHER AMINO ACID SURFACTANTS

Derivatives of lower amino acids reported in the patent literature include p-tert octylphenoxy acetyl glycine (228). Other amino acid surfactants (examples follow) have been proposed in recent years, but their commercial penetration has been hindered by their high prices.

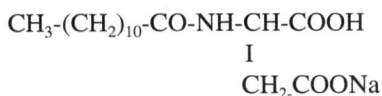

$$CH_3-(CH_2)_{10}-CO-NH-CH-COOH$$
$$I$$
$$CH_2.COONa$$

Sodium lauroyl aspartate.

R-CO-NH-CH-COONa
I
CH₃

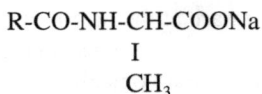

Sodium cocoyl alaninate.

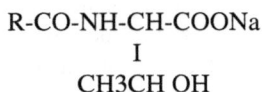

R-CO-NH-CH-COONa
I
CH3CH OH

Sodium cocoyl threoninate.

6.18.6 Hydrolysed Proteins Surfactants

Mixtures of amino acids usually obtained by the acylation of waste proteins hydrolysates (e.g., keratin – corn, hoof, wool – or collagen from meat processing) were used already in the 1930s to produce surface active agents (229, 230, 231, 232). Some were commercialised under the trade name of Lamepon. Products in this category and have found applications in the textile industry as well as in personal care because of their mildness and softening properties. Lamepon S (potassium cocoyl hydrolysed collagen) is a high-foaming, low-cost mild cleanser that is still used in personal care formulations outside "animal-free" Europe. Protein hydrolysates of vegetable origin are today offered by Croda (oat), Seppic (apple), Lonza (soy), and Seiwa Kasei (silk).

Many other variants are reported in the patent literature, for example, the monobenzylated or otherwise aralkylated amino acids or peptides (233, 234) alkylated arylcarboxy acid chlorides condensed with sarcosine or protein hydrolysates (235, 236, 237), and condensation products of fatty amines with one carboxy group of a dicarboxylic acid (238, 239, 240). Intermediate urea and urethane linkages have also been described (241, 242, 243), but these products remained at the stage of laboratory curiosities.

Although the hydrolysed protein surfactants may claim to be sustainable and renewable, thus offering personal care formulators an alternative to petroleum-derived surfactants when the cost performance justifies, they suffer from severe drawbacks:

- The presence of proteins stresses the preservative system and needs its own preservation.
- They are unsuited for acidic formulations.
- Salt and soap content can be high.
- They are available only in diluted form.
- Odour and colour control can be problematic.

BIBLIOGRAPHY

(49) Lebnes Farbes Zeitung, No. 23 (1902)
(50) Bull. Soc. Ind. Mulhouse, page 255 (1909)
(51) Procter, *Principles of Leather Manufacture*, page 441 (1922)
(52) Deutsches Reich Patent 113,433
(53) Deutsches Reich Patent 260 748 (1911)
(54) Grün, Waldenburg, "Reactions of Castor Oil and Ricinoleic Acid with Concentrated Sulphuric and Chlorosulphonic Acid" *Journal of the American Chemical Society* 31, page 490 (1909)
(55) Hildtitch, *Sulphated Oils and Allied Products*, London (1939)
(56) Sisley, J.P., "Sulphonation-Sulphation" *American Dyestuff Reporter*, 43(23) (November 8, 1954)
(57) French Pat. 721,070
(58) US Pats. 1,823,815; 1,974,007; 2,032,313–14
(59) Ger. Pats 625,637; 633,082; 655,942; 659,528; 676,343; 681,441
(60) Thieme, J. Prakt. Chem., 85, 284; Koninkl. Akadem. Wetenschap, Amsterdam, 10, 855 (1909); Chem. Abstracts 4, 754 (1910)
(61) US Pats. 2,023,387; 2,023,388
(62) Ger. Pats. 689,511; 702,598
(63) Ger. Pats. 595,173; 628,828; 671,085; 678,031
(64) US Pat. 1,918,363
(65) French Pat. 718,393
(66) US Pats. 1,918,373; 1,981,792
(67) US Pat. 2,355,503
(68) Can. Pat. 427,726
(69) Brit. Pat. 523,466
(70) US Pat. 2,167,931
(71) Baird, W., *Textile Auxiliary Products Manufactured by I.G. Farben at Ludwigshafen*, PB 28754, Office of Technical Services, Dept. of Commerce, Washington, DC.
(72) US Pat. 2,353,081
(73) Brit. Pat. 414,403
(74) Brit. Pats. 343,899; 341,053
(75) French Pat. 716,178
(76) Brit. Pat. 343,872
(77) Ger. Pat. 705,179
(78) Dumas and Peligot, Ann. Pharmazie 19, page 293 (1836)
(79) Ger. Pat. 705,357
(80) French Pat. 842,184
(81) Brit. Pats 443,559; 443,631; 443,632
(82) US. Pat. 2,203,883
(83) Collegium 1933, page 580
(84) French Pat. 776,044
(85) US Pat. 1,933,431
(86) Ger. Pats 546,897; 593,709
(87) French Pats. 671,456; 751,652; 753,055
(88) US Pats. 1,968,793/97
(89) French Pats. 693,814; 679,186; 735,235

(90)   Ger. Pats. 571,124; 577,220
(91)   Ger. Pat. 608,692
(92)   Baird, *Miscellaneous Surface-Active Agents and Related Intermediates*, British Intelligence Objective Sub-Committee; Final Report Nr. 1151
(93)   Ger. Pat. 336,558
(94)   US Pats. 1,980,543; 2,020,385; 2,133,282; 2,143,493
(95)   Ger. Pat. 544,489
(96)   US Pats. 1,737,792; 1,836,588; 1,901,507
(97)   Dee, F., Snell, C.T. (August 1958). "50th Anniversary Feature-Fifty Years of Detergent Progress" *Industrial & Engineering Chemistry* 50(8), pages 48A–51A, (1980)
(98)   Kocal, J.A., Vora, B.V., Imai, T., "Production of Linear Alkylbenzenes" *Applied Catalysis A: General* 221, pages 295–301 (2001)
(99)   Krafft, Ber. Dtsch. Chem. Ges. 19,1903 (1886)
(100) Adam, Proceed. Roy. Soc. London, Ser. A. 103, 676 (1923)
(101) Ger. Pats. 553,811; 561,603; 567,361
(102) Ger. Pats. 438,061; 583,686
(103) French Pats. 640,617; 679,187; 706,131
(104) Spanish Pat. 454,183
(105) French Pats. 766,903; 795,429
(106) Ger. Pats. 514,080; 544,889
(107) French Pat. 322,005
(108) US pat. 1,980,543
(109) French Pat. 766,903
(110) US Pats. 2,220,099; 2,161,173; 2,223,364; 2,283,199; 2,390,295; 2,394,851; 2,364,767; 2,210,962; 2,072,061
(111) US Pat. 2,385,303
(112) Hoyt, *PB3868*, Offices of Technical Services, Dept. of Commerce, Washington, DC.
(113) US Pat. 2,072,153
(114) Hoyt, *Synthetic Detergents and Washing Agents*, British Intelligence Objectives Sub-Committee (BIOS), Miscellaneous Report Nr. 11
(115) US Pats. 2,081,075; 2,302,070
(116) Ger. Pats. 583,686; 611,443
(117) French Pats. 809,342, 782,612
(118) Foster, D., Cornelia, S.T., "50th Anniversary Feature (9558). Fifty Years of Detergent Progress" *Industrial and Engineering Chemistry* 50(8), pages 48A–51A (1958)
(119) Scheibel, J., "The Evolution of Anionic Surfactant Technology to Meet the Requirements of the Laundry Detergent Industry" *Journal of Surfactants and Detergents* 7(4), pages 319–328 (2015)
(120) Hashim, M.A., Kulandai, J., Hassan, R.S., "Biodegradability of Branched Alkylbenzene Sulphonates" *Journal of Chemical Technology & Biotechnology* 54(3), pages 207–214 (2007)
(121) McKinney, R.E., "Syndets and Waste Disposal" *Sewage and Industrial Wastes* 29(6), pages 654–666 (1957)
(122) Sawyer, C.N., Ryckman, D.W., "Anionic Synthetic Detergents and Water Supply Problems" *American Water Works Association* 49(4), pages 480–490 (1957)
(123) US 2,170,809
(124) Strecker, Liebigs Annal. Chem. 146, page 90 (1868)
(125) Haemilian, Liebigs Annal. Chem. 168, page 145 (1873)
(126) Norris, J. Chem. Soc. (London) 121, page 2161 (1922)
(127) Reychler, A., "Beiträge zur Kentniss der Seifen" *Kolloid Z.* 12, pages 277–283 (1913)

(128)   US Pat. 2,170,380
(129)   French Pat. 716,715
(130)   Brit. Pat. 433,312
(131)   Ger. Pat. 687,462
(132)   French. Pat. 785,561
(133)   Spanish Pat. 353,475
(134)   US Pats. 1,949,837; 1,851,102
(135)   Ger. Pat. 608,831
(136)   French Pat. 721,794
(137)   US Pat. 2,043,476
(138)   US Pat. 2,308,841
(139)   US Pat. 2,398,426
(140)   Wagner and Reid, J. Amer. Chem. Soc. 53, page 3407 (1931)
(141)   US Pats. 2,204210; 2,142,162; 2,187,338/39; 2,338,830; 2,402,587
(142)   US Pats. 2,046,090; 2,174,110; 2,263,312
(143)   French Pat. 849,393
(144)   Ger. Pats. 711,822; 715,323; 721,722; 717,680; 719,059; 721,892; 710,965
(145)   US Pats. 1,765,601; 2,173,133; 2,225,960; 2,142,934; 2,146,744; 2,147,346; 2,174,856;
        2,046,090; 2,174,110; 2,174,111; 2,174,492; 2,242,086
(146)   French Pats. 849,393; 848,001/02; 853,686; 870,163; 870,294
(147)   Spanish Pats. 508,801; 508,794
(148)   Italian. Pats 368,810; 369,149
(149)   Holland Pat. 51,039
(150)   Belgian Pat. 434,141
(151)   Swiss Pat. 210,199
(152)   Hoyt, PB 3868, Office of Technical Services, Dept. of Commerce, Washington, DC.
(153)   Ger. Pat. 735,096
(154)   Ger. Pat. Appl. J69.611; J 69,375; J 74,599
(155)   Belgian Pats. 443,658; 444,546; 445,312; 445,349
(156)   Ger. Pat. 735,096
(157)   US Pat. 3,769,332A (1960)
(158)   US Pats. 2,195, 088; 2,195,145. 2,195, 186; 2,195,188
(159)   French Pat. 785,561
(160)   US Pat. 2,061,617
(161)   AOCS Press, *Biobased Surfactants*, 2nd ed., Synthesis: Properties and Applications,
        pages 303–324 (2019)
(162)   US Pats. 5,723,433; 5,587,500; 6,058,623
(163)   US Pat. 6,657071
(164)   US Pat. Appl. USSN61/026,174
(165)   Mazzani, C., *Introduction of Surfactants from Biorenewable Resources*, AOCS, Nr.5
        pages 1–9, March 2008
(166)   US Pats. 2,289,391; 2,342,563
(167)   French Pat. 788,748
(168)   Ger. Pat. 731,393
(169)   US Pats. 1,881,172; 1,916,776
(170)   Ger. Pats. 652,410; 655,999; 657,357; 657,404; 679,186
(171)   French Pats. 693,620; 720,590
(172)   Brit. Pats. 359,893; 366,916, 372,005
(173)   US Pats. 2,028,091; 2,176,423
(174)   Brit. Pat. 446,568

(175)   French Pat. 776,495
(176)   US Pat. 2,236,528
(177)   US Pat. 2,184,794
(178)   US Pat. 2,345,041
(179)   Brit. Pat. 551,246
(180)   US Pats. Pat 1,931,257; 2,348,200; 2,376,381
(181)   Ger. Pat. 545,968
(182)   French Pat. 789,993
(183)   Canadian Pat. 426,101
(184)   Thomas, C.A., *Anhydrous Aluminium Chloride in Organic Chemistry*, Reinhold, New York (1941)
(185)   Schwartz, A.M., Peery, J.W., *Surface Active Agents their Chemistry and Technology*, Interscience Publishers, Inc., New York, page 112 (1949)
(186)   Schwartz, A.M., ibid, pages 112–113
(187)   Hammet, *Organic Chemistry*, chapter 10, McGraw-Hill, New York, 1940
(188)   Nightingale, Chem. Revs., 25, page 320 (1939)
(189)   Pauling, L., *Nature of the Chemical Bond*, Cornell Univ. Press, Ithaca (1940)
(190)   Hoyt, *PB3868*, Offices of Technical Services, Dept. of Commerce, Washington, DC.
(191)   US Pats. 2,081,075; 2,302,070 (to Groggins et al.)
(192)   Ger. Pats. 583,686; 611,433
(193)   French Pats 809,342; 782,612
(194)   *Linear alkylbenzene*, 07/08-S7 Report, Chem Systems, February 2009, Archived July 8, 2011, at the Wayback Machine
(195)   Petroianu, G., A. Pharmazie 2010, Oct. 65 (10), pages 776–780
(196)   Turowski, A., Skrypzak, W., Miller, D., Weilman, Ch., "Alkylphosphorsäureester- Ein Beitrag zu nicht ethoxylierten Emulgatoren" *Vortrag gehalten auf dem* 44. SEPAWA- Kongress 1997 in Bad Dürkheim
(197)   Paul, S.L., "Preparation and Industrial Applications of Surfactants" in Karsa, D.R., ed. *Industrial Applications of Surfactants II*, Special Publication No. 77, The Royal Chemical Society, Thomas Graham House, Cambridge, UK, pages 115–131 (1989)
(198)   Aigner, R., Loeffler, A., Overweg, A. and Turowski, A., "Alkyl Phosphates as Cosmetic Emulsifiers; A Contribution to Non-ethoxylated Emulsifiers" in Karsa, D.R., ed. *Industrial Applications of Surfactants IV*, Special Publication No. 230, The Royal Chemical Society, Thomas Graham House, Science Park, Milton Road, Cambridge, UK, pages 227–243 (1989)
(199)   Imokawa, G., "Comparative Study on the Mechanism of Irritation by Sulphate and Phosphate Type of Anionic Surfactants" *Journal of the Society of Cosmetic Chemists* 31, pages 45–66 (March/April 1980)
(200)   Solvay Soprophor® and Innospec Enviomet® brochures
(201)   Dario, B.S., Pereira, R., Petri, D.F.S., et al., "Tristyryl Phenol Based Surfactants as Efficient Dispersants of TiO2 in Diluted and Concentrated Dispersions" *Elsevier, Colloids and Surfaces A: Physicochemical and Engineering Aspects*, 654, December 2022
(202)   French. Pat. 851, 842
(203)   Ohl, Monatsch. Seide u. Kunstliche Zelfolle, 43, 411 (1938)
(204)   Vorobev, Chemie & Industrie, 32, 1405 (1938)
(205)   US 2.384.033
(206)   US 2.176.510
(207)   Plasticizers Information Center
(208)   Karsa, D.R., Private Communication

(209)   Karsa, D.R., Private Communication
(210)   Karsa, D.R., "Industrial Applications of Surfactants, in Industrial Applications of Surfactants I" in Karsa D.R., ed., Royal Society of Chemistry, Industrial Division N.W. Region III Series, pages 17–20 (1987)
(211)   Paul, S.L., "Preparation and Industrial Applications of Surfactants", in Karsa, D.R., ed., *Industrial Applications of Surfactants II,* Special Publication No. 77, The Royal Chemical Society, Thomas Graham House, Cambridge, UK, pages 115–131 (1989)
(212)   US Pat. 1,973,860
(213)   Pötsch, W., *Lexikon bedeutender Chemiker,* VEB Bibliographisches Institut, Leipzig, (1989)
(214)   US Pat. 1,939,180
(215)   Ger. Pats. 584,73
(216)   Brit. Pats 341,053; 343,524; 36,982; 372,389; 389,543
(217)   US Pat. 2.888.219
(218)   Burnette, L.W., Chiddix, M.H., "Reaction of Fatty Acids with N-Methyl Taurine" *Journal of the American Oil Chemists' Society* 39(11) (1962)
(219)   French Pat. 721,998
(220)   Brit. Pat. 446,912
(221)   Brit. Pat. 434,358
(222)   Ger. Pat. 677,600
(223)   Swiss Pat. 210,962
(224)   Ger. Pat. 635,522
(225)   Brit. Pats. 456,142; 459,039; 461,328
(226)   Takehara, M., Yoshimura, I., Takizawa, K., Yoshida, R., "Surface Active N-Acylglutamate: I Preparation of Long-chain N-Acylglutamic Acid" *Journal of the American Oil Chemists' Society* 43(2) (March 1972)
(227)   US Pat. 9 626 787 B2
(228)   US Pat. 2,215,367
(229)   US Pats. 2,015,912; 2,041,265
(230)   Ger. Pats. 692,070; 697,324
(231)   Brits Pats. 413,016; 435,481; 450,467
(232)   French Pats. 772,585; 772,636
(233)   US Pat. 2,119,872
(234)   Ger Pats. 670,096; 670,097
(235)   US Pat. 2,215,367
(236)   Brit Pat. 498,136
(237)   French Pat. 839,600
(238)   US Pat. 2,191,738
(239)   French Pat. 795,662
(240)   Brit. Pat. 510,308
(241)   Brit. Pat. 510,310
(242)   US Pats. 2,143,49; 2,157,362; 2,251,892
(243)   Ger. Pat. 585,161

7 Cationic Surfactants

7.1 INTRODUCTION

Cationic surfactants are those that dissociate in water giving an amphiphilic cation and an anion, most often of the halogen type. A large proportion of this class corresponds to nitrogen compounds, such as fatty amine salts and quaternary ammonium with one or several long-chain alkyl types, mostly coming from the transformation of fatty acids.

The industrial development of fatty amines chemistry, on which the cationic surfactants are largely based, started just after World War II. It is an example of highly diversified and sophisticated technology that, however, relies on largely available and inexpensive raw materials like animal fats and vegetable oils.

The main markets for cationic surfactants were originally in industrial disinfection, pharmaceutical, anti-corrosion, pigment coating, and, later, hair rinses/conditioners and domestic fabric softeners. The quaternary salts (abbreviated in "quats") of mono alkyl and alkyl aryl trimethyl amines and dialkyl dimethyl amines were used.

In the mid-1980s, a campaign over the poor quats biodegradability and the possible bioaccumulation in the environment of the ditallow dimethyl ammonium chloride (DTDMAC) (the key ingredient in domestic fabric softeners) led to their (almost overnight) total replacement by the ester quats in Europe and in many formulations in the US.

7.2 CATIONIC SURFACTANTS PRODUCTION PROCESS AND RAW MATERIALS

7.2.1 QUATERNARY AMMONIUM COMPOUNDS

All surface active quaternary ammonium compounds are prepared by either:

- Exhaustive alkylation of a primary or secondary fatty amine
- Alkylation of a tertiary fatty amine
- Alkylation of a low-molecular-weight tertiary amine such as trimethyl amine, trethylamine, or pyridine with a fatty alkyl halide
- Treatment a tertiary amine or its salt with an epoxide

The large majority of the quaternary ammonium compounds is produced via the exhaustive alkylation of primary or secondary amine using methyl chloride

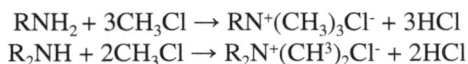

$$RNH_2 + 3CH_3Cl \rightarrow RN^+(CH_3)_3Cl^- + 3HCl$$
$$R_2NH + 2CH_3Cl \rightarrow R_2N^+(CH^3)_2Cl^- + 2HCl$$

The reaction will not go to completion in an acid medium, and in the industrial process, sodium or potassium carbonate is used to scavenge the HCl produced during

DOI: 10.1201/9781003403869-9

the reaction. The reaction is carried out in a polar solvent (usually water) at 60–95 °C and slight pressure (244).

The principal raw material to produce quaternary ammonium compounds is a fatty mono- or di-alkyl amine. For the production of mono-alkyl amines, I.G. Farbenindustrie A.G. developed a process using fatty acids as the starting raw material (245, 246). The process involves the reaction of a fatty acid with ammonia to form an amide, which, in turn, is dehydrated to nitrile:

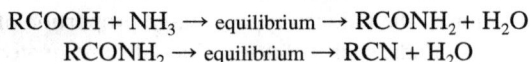

$$RCOOH + NH_3 \rightarrow \text{equilibrium} \rightarrow RCONH_2 + H_2O$$
$$RCONH_2 \rightarrow \text{equilibrium} \rightarrow RCN + H_2O$$

To optimise the yield in nitrile the reaction conditions have to be carefully balanced (247). The preheated fatty acid is continuously counter-current reacted with ammonia at 280–330 °C and a pressure up to 100 psi for up to 3 hours. Water and ammonia are removed at the top, and the excess ammonia is recovered. The mixture of nitrile, amide, and fatty acid at the bottom is vapourised with more ammonia in the presence of a bauxite catalyst, causing the almost total conversion to the nitrile.

In an improved continuous process (248), fatty acids and ammonia are brought together in the first reaction zone. The mixture of amide and fatty acid is reacted with more ammonia in a second zone; in a third zone, the amide and nitrile are heated with more ammonia to yield an essentially pure nitrile. The reaction is carried out at 300 °C under sufficient pressure to keep all materials liquid. The nitriles are then converted to primary amines by hydrogenation with a Raney Nickel catalyst. The reaction is complex and involves several competing reaction steps:

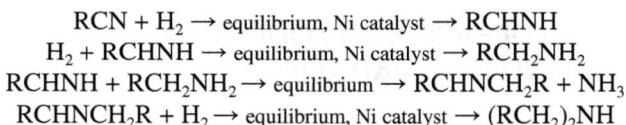

$$RCN + H_2 \rightarrow \text{equilibrium, Ni catalyst} \rightarrow RCHNH$$
$$H_2 + RCHNH \rightarrow \text{equilibrium, Ni catalyst} \rightarrow RCH_2NH_2$$
$$RCHNH + RCH_2NH_2 \rightarrow \text{equilibrium} \rightarrow RCHNCH_2R + NH_3$$
$$RCHNCH_2R + H_2 \rightarrow \text{equilibrium, Ni catalyst} \rightarrow (RCH_2)_2NH$$

To maximise the yield in primary amines, ammonia is introduced or sodium or potassium hydroxide is used instead of ammonia to force the equilibrium reaction to the left.

Primary fatty amines can be obtained also by the ammonolysis of fatty alcohols. Fatty alcohols and ammonia are heated at 380–400 °C at 120–130 Atm in the presence of an Al_2O_3 catalyst (249). The reaction can be performed with a variety of alcohols and primary or secondary amines instead of ammonia (250).

The reductive amination of fatty alcohols was industrially practised since mid-1970 by Imperial Chemical Industries to produce their Synprolam range of synthetic C13–15 primary amines.

$$RCH_2OH \rightarrow -H_2, \text{catalyst} \rightarrow RCHO + H_2 + NH_3 \rightarrow RCH_2NH_2$$

Other methods include the amination of α-olefins (251), the ammonolysis of hydrogenated hydrocarbons, and the reduction of cyanamides (252).

Secondary amines are obtained by the hydrogenation of fatty nitriles under reaction conditions that continuously eliminate ammonia, which inhibits the formation off secondary amines. However, the industrial production of secondary amine is carried out more economically in a two-step process. The nitrile is first converted at low temperature to amine, which is then de-ammoniated to secondary amine using a copper chromite catalyst (253).

Tertiary fatty alkyl amines were synthesised from fatty alcohols and dimethyl amine with a high yield and selectivity using a Cu/Ni/Ca/Ba catalyst (254). The patents issued in the 1930s describe the preparation of tertiary amines by refluxing an alkyl halogenide with a lower amine (255).

7.2.2 Ester Quats

Ester quats were first synthesised in the 1930s, but at the time and even later (until the mid-1980s), they did not find commercial applications.

The patent literature describes many molecular structures, among those related to fabric softening (256) focused on the feature of ready biodegradability, but two products are now almost universally used:

- The diester of methyl diethanolamine (MDEA) with C16-C18 fatty acids, quaternised with methyl chloride or dimethyl sulphate (DEEDMAC)
- The diester of triethanolamine (TEA) with the previously mentioned fatty acids and quaternised preferentially with dimethyl sulphate (TEAQ)

The production technology of ester quats is comparatively simpler than that of DTDMAC as it just involves the following:

- The hydrolysis of oils/fats and the purification of the fatty acids
- An esterification step of an alkanolamine with a stoichiometric amount of fatty acids of suitable quality
- A quaternisation step

In practice, the manufacturing processes are different, depending on the structure and raw materials.

The complete synthesis path of DEEDMAC follows:

- Tallow or palm oil → glycerin + fatty acid → distilled fatty acids
- 2 moles (nominal) of fatty acid + 1 mole MDEA + catalyst → water + methylamine diester
- Methylamine diester + methyl chloride + ethanol solvent → DEEDMAC

The fatty acid is left unsaturated as in soft tallow, because the final ester quat is lower melting and easier to formulate. To improve the colour and odour stability of the final quat, the fatty acid may undergo additional processing, such as mild hydrogenation or clay treatment. This adds costs, time in production, and the need for an additional reactor. The quaternisation with methyl chloride requires the installation of methyl chloride recovery, which is tricky and expensive.

The complete synthesis path of TEAQ follows:

- Tallow or palm oil → glycerin + fatty acid → distilled fatty acids
- 2 moles of distilled fatty acids + TEA + catalyst → water + trimethylamine diester (in reality a mixture of mono-, di-, and triesters)
- Trimethylamine diester + dimethyl sulphate + isopropanol as solvent → TEAQ

The production of TEAQ is less capital-intensive than the DEEDMAC as only three reactors are needed (the last two process steps can be done in the same multipurpose reactor). If fatty acids are purchased, only one multipurpose reactor equipped with an agitator and a condenser is sufficient. The condenser will allow the esterification water to leave and retain TEA; during the quaternisation, the condenser will keep iso-propyl amine in the reactor. It is recommended to have nitrogen sparging to facilitate water removal and preserve colour. The application of a vacuum towards the end of the esterification speeds up the final water removal. Nitrogen blanketing is needed for fatty acids and TEAQ storage.

BIBLIOGRAPHY

(244)   US Pats. 2,950,318; 2,775,617; 3,175,008
(245)   French Pat. 785,622
(246)   Saphiro, S.H., *Fatty Acids and Their Industrial Applications*, E.S. Pattison, ed., Marcel Dekker, New York, pages 109–19 (1968)
(247)   US Pats. 2,314,894; 2,414,393; 2448,275; 2,504,045; 2,524,831; 2,808,426
(248)   US Pat. 3,299,117
(249)   Ger. Pat. 611.924
(250)   Linfield, W.M., *Cationic Surfactants*, Jungerman, E., ed., Marcel Dekker Inc., page 22 (1970)
(251)   Brit. Pat. 339.962, equivalent to Swiss Pat. 150.000 and addition Swiss Pats. 153.089 to 153.091
(252)   Huan, W., Xien, L., Jibo, Z., "New Materials and Intelligent Manufacturing (NMIM)" *Topics in Chemical and Material Engineering* 1, pages 191–194 (2018)
(253)   US Pat. 2,355,356
(254)   Hiroshi, K., Hideki, T., "Targeting Quantitative Synthesis for the One-Step Amination of Fatty Alcohols and Dimethyl Amine" *Applied Catalyst A: General* 287(2), pages 191–196 (2005)
(255)   French Pat. 696.328, equivalent to US Pats.1.836.047; 1.836.048
(256)   US Pats. 3,915,867A; 4,370,272A

8 Nonionic Surfactants

8.1 INTRODUCTION

Nonionic surfactants are those that do not dissociate in water to give ionic moieties. Within the scope of this definition, the following should be classified as nonionic surfactants:

- Alkoxylated fatty alcohols and alkylphenols
- Alkoxylated fatty acids
- Alkoxylated fatty amines
- Alkoxylated fatty amides
- Alkoxylated mercaptans
- Fatty acid esters of polyhydric alcohols, for example, glycerol, sorbitol/sorbitan, neopolyols, sugar, and their alkoxylated derivatives
- Alkyl polyglucosides
- Amine oxides
- Esters of fatty acids with monohydric and polyhydric alcohols, for example, lower alcohol, glycols, glycerol, sorbitol/sorbitan, neopolyols, and sugar
- Ethoxylated methyl esters
- Ethylene oxide (EO)/propylene oxide (PO) block and random copolymers (lower alcohols or glycols or fatty alcohols initiated)
- Fatty acid esters of alkoxylated mono- and polyhydric alcohols
- Polyglycerol esters

Sometimes, however, fatty amines and fatty amides alkoxylated, and the amine oxide are classified as cationic because of the polarisation of the nitrogen–oxygen bonding in the amine oxide and the slight cationic character (as measured by their pKa) of the alkoxylated amines. This classification is not retained in this book.

8.2 ALKOXYLATION OF FATTY ALCOHOLS AND ALKYL PHENOLS

The oxyethylation of alcohols was first reported in 1926, but it was the work from Schoeller and Wittwer first (257) and from Steindorff, Bolle, Horst and Michel (258) that, in the 1930s, led to the development and patenting of the synthesis of ethoxylated nonionics from fatty acids, alcohols, amines, and alkylphenols.

The patent literature on the oxyethylene derivatives of substrates containing active hydrogen is extensive. Other than the two earlier references, it is worth reporting the following:

- French Pats. 723,426; 717,427; 727,202; 752,831; 770,804
- Ger. Pats 548,201; 670,419; 680,245

DOI: 10.1201/9781003403869-10

- Brit. Pats. 346,550; 367,420; 380,431; 409,336; 443,559; 443,631
- US Pats. 2,213,417; 2,214,352

The combined use of EO and PO was first proposed to modify the physical-chemical properties of fatty moieties, in particular water solubility (259).

Of future significant commercial importance was the prior art of Belgian Pat. 444,625, which described the oxyethylation of higher fatty alcohols made by the reaction of olefins, carbon monoxide, and hydrogen. These surfactants became popular alternatives to those obtained from natural alcohols once the hydroformylation process from Shell (SHOP), Exxon, and ICI made available to the market significant volumes of synthetic fatty alcohols.

The ethoxylation of alkylphenols yields a class of nonionic surfactants that quickly gained widespread acceptance because of their flexibility and cost-effectiveness and attained significant consumption volumes.

Alkylphenols with alkyl chains C6 to C14 were produced by the alkylation of phenol with olefins catalysed by hydrofluoric acid or boron trifluoride. The alkylating olefins were blends of hexane/heptane and dodecene (obtained through dimerisation of hexene/heptene, trimerisation of butylene, or tetramerisation of propylene).

Alkyl naphthols were easily prepared with the same catalysts and alkylating agents.

Phenols, alkyl phenols, and their derivatives have always been regarded with suspicion. Phenol is highly toxic and has a lethal dose as low as 50 mg/kg; severe systemic toxicity occurs after any route of exposure, with major damage to the liver, kidneys, and eyes, and it is highly caustic to tissues (260). No surprise that this cast suspicion on the toxicology of its derivatives as well.

The toxicological properties, biodegradability, and aquatic toxicity of nonyl phenol ethoxylates (NPEs) depend on the chain length of the ethoxylated chains. In general terms, acute and sub-acute oral toxicity is low. The lowest observed adverse effect level measured in rats is of 43–400 mg/day (261).

NPEs do not meet the criteria for classification as "readily biodegradable". They are moderately toxic to toxic to fish and aquatic organisms on an acute basis. In the environment, NPEs degrade to the more toxic and persistent nonyl phenol. In the 1990s, because of the endocrine effects in fish caused by the metabolites and NPEs, these are classified hormone-disrupting substances (262).

In the EU, the use of NPEs has been banned in all products leading to a discharge in rivers and lakes. In the US, the use of NPE is prohibited in some states, while the Environmental Protection Agency (EPA) and detergent manufacturers, in a joint action, have cooperated to eliminate their use. However, NPEs are still present in industrial and institutional laundry and cleaning formulations and other uses that lead to discharge in the environment.

8.3 FATTY ACID ESTERS OF POLYHYDRIC ALCOHOLS AND THEIR ALKOXYLATED DERIVATIVES

The sorbitan esters of C12–C18 fatty acids (trade name "Span") and their ethoxylates derivatives (trade name "Tween") deserve special consideration, as the products, and the associated concept of the hydrophilic/lipophilic balance (HLB; originally developed to promote them commercially), provided a tremendous contribution to spreading an understanding and practical implementation of the practice of emulsion making (263, 264). The personal care and pharmaceutical industries were the first and largest beneficiaries of these developments that, however, with the years, extended to other industrial segments and applications.

8.3.1 THE HLB CONCEPT

The HLB concept is reported here just because of its contribution to the early developments of formulation technology. With the advances in colloid chemistry and applications, it is now out of date, as well as its supposed improvement (the Davies numbers). Recent attempts at a novel approach, the Hidrophylic-Lipophilic-Distance (the HLD theory), although conceptually captivating, have still to fully confirm the expectations.

In the late 1940s and early 1950s, W. Griffin from the Atlas Chemical Company introduced the Hydrophilic Lipophilic Banace (HLB) concept to categorise the nonionic surfactants and assist in the choice of emulsifiers or combination of emulsifiers to optimise the emulsification process and the emulsion quality.

In the Griffin concept, the HLB of a nonionic surfactant can be expressed as

$$HLBGriffin = 20*(Mh/M)$$

where Mh is the molecular mass of the hydrophilic portion of the molecule and M is the molecular mass of the entire molecule. This gives a scale of 0–20: 0 corresponds to a completely hydrophobic molecule and 20 to a completely hydrophilic (263, 264).

The preceding equation is valid for ethoxylated surfactants. For esters, the following formula (Figure 8.1) was proposed.

The Griffin HLB is of practical interest only for comparing and tentatively predicting the performance of ethoxylated surfactants. Surfactants such as anionic and cationic surfactants or compounds with complex structures cannot be described by the preceding equations. In 1957, Davies suggested a method based on calculating an HLB value based on the chemical groups of the molecule (265). The advantage of this method is that it considers the effect of stronger and weaker hydrophilic groups and that in principle, it should be applicable to all surfactant classes. The method works as shown in Figure 8.2.

$$\textbf{HLB}_{Griffin} = 20 \left(1 - \frac{SV}{AV} \right)$$

FIGURE 8.1 Griffin's HLB of esters, where SV is the saponification value of the ester and AV is the acid value of fatty acid.

Sources: From References 263 and 264.

$$\mathbf{HLB}_{Davies} = 7 + \sum H_{h,i} - \sum H_{l,j}$$

FIGURE.8.2 The Davies HLB, where $H_{h,i}$ correspond to the hydrophilic part and $H_{l,j}$ are those of the lipophilic part.

Source: From Reference 265.

TABLE 8.1
The Davies Number

Hydrophilic Group	Group Number
$-SO_4^- Na^+$	38.7
$-COO^-K^+$	21.1
$-COO^-Na^+$	19.1
-COOH	2.1
N (tertiary amine)	9.4
Ester (sorbitan ring)	6.8
Ester (free)	2.4
Hydroxyl (free)	1.9
Hydroxyl (sorbitan ring)	0.5
-O-	1.3

Lipophilic Groups	Group Number
-CH-	-0.475
$-CH_2-$	-0.475
CH_3-	-0.475
=CH-	-0.475

The values of the common chemical groups are given in Table 8.1.

The original calculation of the Davies HLB number has been refined in recent years (266, 267), but it remains essentially of academic interest and is no longer used.

In a further attempt to overcome the shortcomings of Griffin's HLB in the years, between 2015 and 2019, Prof. Steven Abbot developed the concept of the HLD. This new approach originated from the demand for surfactant formulations offering cost/effective options for enhanced oil recovery and retains this imprint in its expression. The objective is to balance a whole system of water, surfactant, oil type, temperature, and salt. It is too early to express a judgement on the real value of the HLD concept, which is reported for the sake of completeness.

8.4 SORBITAN ESTERS

The esterification of a liquid solution of sorbitol with fatty acids leads to a variety of mild surface active agents commonly known as sorbitan esters. Under esterification conditions and high temperatures, sorbitol undergoes internal dehydration to give a tetrahydrofuran structure (sorbitan) that is esterified by the fatty acid mainly on the primary hydroxyl group. When an excess of fatty acid is used, diesters and triesters are obtained.

The years of the 1930s saw the first reported esterification of sorbitol with fatty acids, which used alkaline cell–liquor sorbitol from the electrochemical process, without added catalysts. Sorbitan esters were further developed in the late 1930s/ early 1940s by Atlas Powder Company in Wilmington, Delaware, as a way to extract value from the large amounts of sorbitol that were a by-product in the production of mannitol, at the time a key raw material in the manufacture of detonators for explosives. The original one-step process described for example by K.R. Brown (268) and W.C. Griffin (269) underwent several improvements that led to a two-stage catalyst addition technique (270). The first step is base catalysed ($Ca(OH)_2$ or Na OH) at around 230–240 °C, which leads to the virtual completion of esterification of the sorbitol by the fatty acid and is followed by an acid-catalysation stage (e.g., H_3PO_4) in which the desired degree of anhydrisation is achieved.

Sorbitan esters are largely used as low-HLB emulsifiers in the food industry, for example in chocolate toppings, as (like most surface active esters) they are generally recognised as safe (GRAS). They are also used in personal care, pharmaceuticals, and in some industrial applications, for example, aluminium rolling, emulsion explosives, and colourant masterbatches.

8.5 POLYSORBATES

Sorbitan esters are lipophilic but can be further reacted with various amounts of EO at the remaining hydroxyl sites resulting in a significant increase in the surfactant's hydrophilicity and water solubility. These products are commonly referred to as polysorbates.

Griffin described the preparation of polyoxyethylene (10) mannitan monostearate in two patents (271). Polysorbates of lauric, myristic, palmitic, stearic and oleic sorbitan mono esters, sesquiesters, and triesters with 4–20 moles of EO were synthesised. Some remained laboratory curiosities, but polysorbate 20 (20 EO sorbitan monolaurate) and 80 (20 EO sorbitan monooleate) achieved large market penetration.

Because of the excellent record of safety in use, polysorbates are largely used as high-HLB emulsifiers (often in combination with sorbitan or glycerol esters) in the personal care and pharmaceutical industries. They are used in food preparations, for example, salad dressings, ice creams, and margarine. However, as polysorbates contain EO moieties, their use in food is more regulated than are the use of sorbitan esters. There are a few industrial applications, for example, liquid colourants for food contact plastics and as co-emulsifiers for lubricants in aluminium rolling.

8.6 ALKYLPOLYGLUCOSIDES

Low molecular weight (MW) alkylpolyglucosides (APGs) were first synthesised by Fisher at the beginning of the 20th century (272). The potential of APGs as surface active agents was recognised as far back as the early 1930s when Th. Boehme obtained a patent on the "Process for the production of glucosides of higher aliphatic alcohols" (273). For mysterious reasons, the patent that claims a process for producing C6–C18 APGs did not attract any attention, and APGs fell into oblivion for the following 30 years, except for a reference to possible industrial applications published in early 1945 that, however, claims only lower alcohols (up to C8) (274).

APGs emerged to the attention of the industry in mid-1960 thanks to Rohm & Haas, that used the process described in US 3,219,656 to produce higher MW APG produced from alcohols of chain length up to C14.

The synthesis of surface active species (higher than C8) was hampered by the process constraints that demand that the large excess of the alcohols that have to be removed once the reaction is completed. This limits the alkyl chain length to C12–C14 at the maximum. The reaction temperature is a balance between the reaction kinetics and limiting the caramelisation of the glucose. In any case, the reaction is long, and the crude product is very dark in colour.

In consequence, for many years, the volumes produced were small, and the only application was high-alkalinity bottle-washing formulations. This exploited the excellent hydrotrope properties of the C8–C10. Because of these properties, in the second half of the 1980s, C8–C10 APGs were chosen to enhance the performance of non-selective herbicides (sulphosates, glyphosates). But until the mid-1980s, only two products were commercially available, Triton BG 10 and Triton CG 110, from Rohm and Haas.

A turning point in APG technology was the production of lightly coloured products by post-reaction purification using a wiped film evaporator or molecular distillation (275). This opened the door for the formulation of C8–C10 and C12–C14 in personal care and hand dishwashing detergents.

Over the years, companies like Procter & Gamble, Henkel and Cie, Colgate-Palmolive, and Uniqema took an interest in the products. Colgate in the US formulated APGs in its hand dishwashing product to exploit their mildness and foaming properties. The experience failed because of cost-effectiveness and a lack of recognition of the mildness by the consumers, and APGs were eventually withdrawn. Henkel and Cie had a similar approach in Europe with its hand dishwashing product and met a similar fate. It tried to download the large capacity that had been built into the household and crop protection markets. ICI Surfactants (later to become Uniqema) focused on crop protection adjuvants. SEPPIC in Europe addressed the personal care sector, introducing C16–C18 APGs as emulsifiers (although only a minor proportion of the product was APG, the balance being unreacted alcohol). They contributed to APG technology by replacing the sulphuric acid catalyst with alkylbenzene sulphonic acid, a surface active agent on its own, that facilitates the phase transfer in the initial stages of the heterogeneous reaction.

Several modifications to APGs have been investigated and products synthesised. These include esters, ethoxylates, sulphates, ethers, carbonates, sulphosuccinates, ether carboxylates, isethionates, and quaternary ammonium compounds (276). To the best of our knowledge, none has received a positive market response.

Today, the main use of APGs remains as adjuvants for non-selective herbicides and as mild cleansing agents and emulsifiers in personal care formulations.

8.7 AMINE OXIDES

Amine oxides were known and studied already before 1900. Short-chain trialkyl amine oxides were isolated in 1909 in marine animals. The mechanism for the formation of amine oxides from tertiary amines using hydrogen peroxide was first proposed by Wieland in 1921 (277) and involves ammonium peroxide (or amine perhydrate) as an intermediate, followed by splitting off water.

$$R_3N + H_2O_2 \rightarrow [R_3NH].[OOH] \rightarrow R_3N \rightarrow O + H_2O$$

The ammonium peroxide intermediates have since been isolated with the subsequent conversion to amine oxide by heat (278), thus confirming the earlier postulate.

Amine oxides are mentioned in Brit. Pat. 437,566 (A) (1935), which describes the amine oxides of trialkylamines containing at least one alkyl from capryl, lauryl, cetyl, stearyl, octadecyl moiety produced by oxidation with peroximonosulphuric acid, hydrogen peroxide, or ozone. The products are reported as having wetting, cleaning, levelling, emulsifying, softening, and direct-dye-fastening properties.

It was not until 1939, with the issuance of an I.G. Farbenindustrie A.G. patent (279) that materials such as dimethylamine oxide were recognised as surfactants. After another 22 years, their utility in liquid household formulations was disclosed, and widespread interest was eventually generated (280).

Excellent reviews of the chemistry and methods of preparation of amine oxide were produced in the early 1960s by Lake and Hoh (281), Lindner (282), and Jungermann and Ginn (283).

US 3,332,999 of 25.07.1967 (to the Ethyl Corporation) describes a process for producing fatty alkyl dimethylamine by ozone oxidation.

Other amine oxides reported for their potential of interest in surfactants applications areas follow (284):

• N-dodecyl-N,N',N'-trimethyl-1,3-proylenediamine-N,N'-dioxide
• N-dodecylmorpholine oxide
• 1-hydroxyethyl-2-octadecylimidazoline oxide
• N,N',N'tris(2-hydroxyethyl)-N-octadecyl-1,3-propylenediamine-N,N'dioxide
• Sodium salt of 2-octadecyl-1-(3-sulphopropyloxyethyl)imidazoline-1-oxide

Amine oxides are highly polar, with the greatest electron density at the oxygen atom. A dipole moment of 4.38 mD has been calculated for the semipolar N→O bond. Depending upon the pH in aqueous solutions, fatty amine oxides behave as nonionic (neutral or alkaline pH) or cationic (acidic pH).

Historically, amine oxides entered the market of surface active agents as foam boosters in hand dishwashing liquid detergents, replacing traditional alkanolamides due to their considerable cost/performance advantage (about one third is required for the same foam-level consistency and permanence) that offset the higher cost. They find use in hair shampoos as foam boosters and a conditioning effect and as foam boosters in body shampoos. Patent literature indicates other possible uses, for example, lime soap dispersants, dye-bath assistants, viscose processing, and foam stabilising in rubber processes.

8.8   ETHOXYLATED FATTY AMINES

The preparation of surface active agents from fatty amines and EO was first described by Schöller and Wittwer and patented in the UK and France (285, 286). Patents in the US and Germany were granted 2 years later (287, 288). In the following years, many patents were issued to several companies in different countries, covering the reactions of EO on primary, secondary, and tertiary amines or amines with active hydrogen (289, 290, 291, 292, 293, 294).

In the reaction of EO with primary amines, both reactive hydrogens of the amine are substituted before further polyoxyethylation occurs (295). The alkylamine is sufficient alkaline to initiate the reaction at around 100 °C. Keeping the molar ratio of amine to EO just below 2 allows obtaining water-clear products, which were used, for example, as antistatic agents in thermoplastic polymers. Higher ethoxylation causes discolouration and needs the addition of a basic catalyst (powder sodium, potassium hydroxide, or sodium methoxide) and a temperature of 150 °C or higher (Imperial Chemical Company Ltd., unpublished results).

Polyoxyethylene fatty amines (POEFAs) have a slight cationic character, but on increasing the degree of ethoxylation, they progressively acquire the properties of nonionic surfactants. They have been used as emulsifiers, foaming agents, corrosion inhibitors, emulsion breakers, mud-drilling additives, levelling agents for dyes, textile finishing agents, antistatic agents, and crop protection adjuvants.

Tallow amines with 15–20 moles of EO have been the most successful products. They were used as adjuvants for non-selective herbicides, notably Monsanto's Roundup® glyphosate. POEFAs allow effective uptake of water-soluble glyphosate across plant cuticles, which are hydrophobic, and reduce the amount of glyphosate washed off plants by rain. The levels of use range from about 1% in tank mixes and up to 21% in the formulated product.

In the golden years of the POEFA, an estimate of 150 ktons/year were used globally just for this application, but recently, the concern over the carcinogenicity of glyphosates and the regulations on the transport and storage of POEFAs have led to a progressive decrease in demand and a phasing out in favour of other adjuvants (e.g., APGs are used in Syngenta's Touchdown® sulphosate formulations).

The POEFA CAS No. 61791–26–2 is on the US EPA List 3 of Inert Ingredients of Pesticides.

8.9   ETHOXYLATED METHYL ESTERS

Ethoxylated methyl esters are a relatively recent development, originally prompted by the potential economic advantages of converting directly fatty acid methyl esters (FAMEs) into ethoxylate derivatives, thus bypassing the hydrogenation step to fatty alcohols.

Much work was done, among others, by Hoechst (296), Henkel (297), Vista (298), and Lion (299), and the products were presented at the 5th World Surfactant Congress (300).

FAMEs cannot be directly ethoxylated with conventional catalysts such as potassium hydroxide, sodium hydroxide, or sodium methoxide, and special catalysts had to be developed (301, 302, 303).

The key features of fatty methyl ester ethoxylates follow:

- Low foaming compared to fatty alcohols ethoxylates of similar degree of ethoxylation and hydrocarbon chain length.
- Due to the special catalysts (304), they have a narrow homologues distribution.
- No gel formation.

There has been a lukewarm interest in the FAMEs, especially from personal care and household detergents, but the difficulties in the manufacturing process and the higher cost compared to conventional fatty alcohols ethoxylates have limited their market penetration.

8.10   EO, PO, AND EO/PO HOMO- AND COPOLYMERS

The polymerisation of EO was first reported by Wurtz (305).

In later works, Staudinger and Schweizer prepared a series of polyethylene glycols (PEGs) from EO and separated the polymers according to their MW (306).

Already in the 1930s, PEGs were produced industrially based on the addition of EO to ethylene glycol under basic conditions. Within one decade, a large variety of applications were developed in areas ranging from cosmetics and pharmaceuticals to lubricants and detergents.

In 1940, Flory established the mechanism of the base-initiated EO polymerisation, predicting a Poisson-type distribution for a living chain-growth process (307).

The anionic polymerisation of PO was developed in the 1940s to generate liquid polyols that found application in hydraulic fluids and lubricants.

One of the first disclosures of a copolymer surfactant is found in US patent 2,174,761 (to I.G. Farbenindustrie A.G.) by H.C. Schouette and M. Wittwer (1939). The initiator was a water-soluble alcohol, such as propyl or amyl or an aliphatic or aromatic compound with 8–18 carbon atoms. In one example, 1 mole of cetyl alcohol is condensed with 4 moles of PO, followed by 20 moles of EO at 140 °C in the presence of a sodium hydroxide catalyst.

A whole range of products were prepared later on. Other than monofunctional alcohols, monofunctional acids, mercaptans, secondary amines, and N-substituted amides were used as initiators. Although polyoxypropylene was preferred as the hydrophobe, polyoxybutylene, polyoxystyrene, and polycyclohexene were also used.

The introduction of EO as a polar, water-soluble block for nonionic block EO–PO–EO copolymers is a development of the 1950s. These copolymers are prepared via the sequential ring-opening polymerisation of PO and EO. First, a central segment of polyoxy PO is synthesised as a precursor and subsequently chain-extended by EO polymerisation. Inverse PO–EO–PO copolymers were synthesised in later years (308).

One of the first recorded patents disclosing the use of the "block" concept indicated that polyoxy propylene glycol could be used as an hydrophobe (309).

EO/PO block copolymers were developed and commercialised by Wyandotte Chemical Corporations in the mid-1950s. They were imitated in the following years by BASF (Pluronics, Tetronics), Dow Chemical Company (Tergitol L series), ICI (Synperonic LF series), and Huntsman (Surfonic POA series). Random copolymers of EO and PO were commercialised by Union Carbide (310). Dow Chemicals released oil-soluble polyalkylene glycols based on PO and butylene oxides (BOs) as performance-enhancing additives in hydrocarbon lubricants (311).

Today, the best-known and most widely used EO/PO derivatives are those produced by adding PO to a propylene glycol initiator, followed by an EO addition (EO–PO–EO). However, there are "reverse" copolymers, where the central backbone is polyethylene oxide (PO–EO–PO), and, of course, the random copolymers (although the reaction kinetics suggest that also in this case, a block structure is ultimately achieved). Other initiators can be used, for example, fatty alcohols giving R–EO–PO or R–PO–EO structures, that have better biodegradability and are used in domestic and industrial automatic dishwashing machine formulations. The ethylene diamine–initiated PO–EOs are excellent pigment dispersants. The nonyl phenol PO–EO is a dispersant for crop protection suspension concentrates.

EO/PO block and random copolymers represent an important class of nonionic surfactants. Due to the high-water solubility of polyoxy ethylene oxide chains in a wide temperature range (0–100 °C) and the low solubility of polyoxy propylene oxide (PPO) chains in water at temperatures exceeding T_c (>15°C; cloud point), these copolymers exhibit an amphiphilic character and surface active properties. They have an excellent toxicological profile and are used whenever possible in cosmetic and pharmaceutical formulations. The low-foaming and anti-foaming properties are widely exploited in many industrial formulations, crop protection, food, paints, inks, pulp and paper, and metalworking, among others.

A remarkable property of the EO–PO–EO block copolymers in aqueous media is their ability to form thermo-reversible gels at a concentration above the critical micelle concentration. This is advantageously exploited in personal care and pharmaceutical formulations.

8.11   ETHOXYLATED MERCAPTANS

Alkoxylated mercaptans were first reported in patents published towards the end of the 1930s (312). But their priority dates to 1934, indicating that work on the chemicals was quite advanced at that date.

The first products were exclusively linear, but later the less-expensive branched-chain mercaptans were synthesised (313, 314). Also, in this case, the priority date of 31.05.1949 indicates an earlier development.

Ethoxylates mercaptans are also claimed in French Pat. 780,144.

These products reportedly have exceptional wetting characteristics over wide water hardness, pH, and temperature ranges and found use as heavy-duty degreasers (e.g., waxes, bitumen) and as metal, electronic, and hard surface cleaners. They were commercialised by GAF (Emulphogene LM series), Rhone-Poulenc (Accodet series), and Monsanto (Siponic SK series).

The toxicology reports serious eye damage; that it is harmful to aquatic life, with long-lasting effects; and that it may cause allergic skin irritation, and for these reasons, their usage has been declining since the 1990s and is confined to a limited range of applications.

8.12   NONIONIC SURFACTANTS PRODUCTION PROCESS AND RAW MATERIALS

Commercial nonionic polyoxyalkylene surfactants are prepared by the addition of a low-molecular-weight epoxide (most commonly ethylene oxide, less commonly PO, or even less commonly butylene oxide) to a hydrophobe containing active hydrogen. The EO addition, which is the one most widely practised can be expressed as

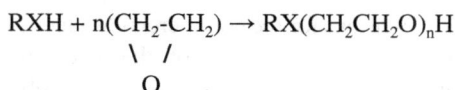

$$RXH + n(CH_2\text{-}CH_2) \rightarrow RX(CH_2CH_2O)_nH$$
$$\backslash \ /$$
$$O$$

but in reality, it is a stepwise series of addition reactions.

The process can be base- or acid-catalysed; however, acid catalysts are rarely used, as they lead to the formation of large amounts of polyethylene glycols and other undesirable by-products. The base-catalysed ring opening of EO is the most important commercial production process.

The ring-opening reactions of EO are nucleophilic additions (315, 316). Under basic conditions, the rate-determining (slow) step involves the attack of the anion (nucleophile) at a carbon atom of the epoxide ring.

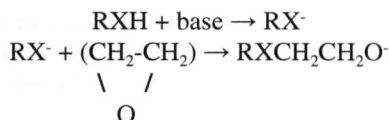

$$RXH + base \rightarrow RX^-$$
$$RX^- + (CH_2\text{-}CH_2) \rightarrow RXCH_2CH_2O^-$$
$$\backslash \ /$$
$$O$$

Since the rate is dependent on $[RX^-][C_2H_4O]$, it is a second-order nucleophilic substitution (SN2). The oxyethylene anion may undergo a fast proton exchange reaction with an additional RXH molecule:

$$RXH + RXCH_2CH_2O \rightarrow RX^- + RXCH_2CH_2OH$$

or may react further with more ethylene oxide:

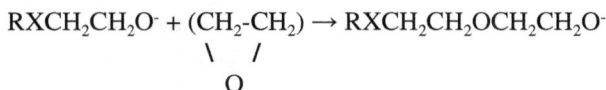

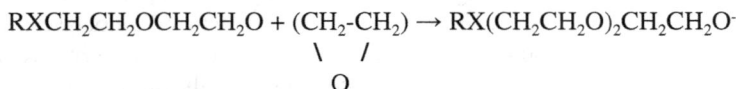

$$RXCH_2CH_2O^- + (CH_2\text{-}CH_2) \rightarrow RXCH_2CH_2OCH_2CH_2O^-$$
$$\backslash \ /$$
$$O$$

$$RXCH_2CH_2OCH_2CH_2O + (CH_2\text{-}CH_2) \rightarrow RX(CH_2CH_2O)_2CH_2CH_2O^-$$
$$\backslash \ /$$
$$O$$
$$\cdots\cdots\cdots\cdots\cdots\cdots\cdots\cdots\cdots\cdots\cdots;$$

$$RX(CH_2CH_2O)_nCH_2CH_2O^- + (CH_2\text{-}CH_2) \rightarrow RX(CH_2CH_2O)_{(n+1)}CH_2CH_2O^-$$
$$\underset{O}{\diagdown\ \diagup}$$

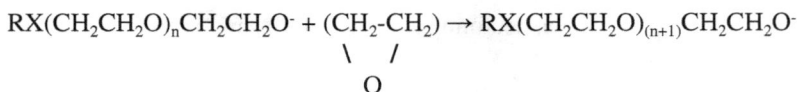

There are no termination reactions. Polymerisation continues until all EO has reacted or the base is destroyed by some other reagent.

The course of the overall reaction is determined by the relative concentrations of RXH and ethylene oxide, the relative reactivities of $RX^- RXCH_2CH_2O^-$ and the relative acidities of RXH and $RXCH_2CH_2OH$. The number of moles "n" reacting with one mole of RXH does not need to be an integer and is the number-average degree of polymerisation of the adducts in the product, each having an integer number n of EO moieties (317).

The process and the conditions for the commercial production of fatty raw-material ethoxylates and EO/PO copolymers depend on the nature of the initiator and the degree of ethoxylation to be achieved, but in essence, it involves three steps.

Preparation of the initiator-catalyst mixture: The initiator and the catalyst are charged in a weight tank, heated and dried. This pretreatment step can be carried out directly in the autoclave, but the use of a separate unit significantly increases plant productivity at limited additional investment costs.

Reaction of the initiator and ethylene oxide: The initiator-catalyst mixture in the autoclave charged with the initiator-catalyst mixture is evacuated and purged with an inert gas (usually nitrogen). The nitrogen pressure is kept at all times, and the partial pressures are adjusted to avoid the explosive limits of the mixture being exceeded. The temperature is raised to initiate the reaction (typically between 120–200 °C, depending on the initiator). EO is fed in the autoclave until the desired pressure is achieved (generally between 1.5–7.0 bar). The initiation of the reaction is signalled by a rise in temperature, at which point cooling is provided. When the reaction is completed (shown as a drop in pressure to a constant level) the batch is cooled and vented.

Removal of catalyst: The basic catalyst is neutralised using organic or inorganic acids. This can be done in the autoclave or in a post-treatment tank (see considerations mentioned earlier). The product may be filtered and de-colourised, depending on the intended use.

Ethoxylation of fatty raw materials is a batch reaction carried out in autoclaves. EO is a flammable, toxic form that is explosive when mixing with air, and it may undergo exothermic decomposition. Over the years, the design of autoclaves has evolved to provide greater and greater safety in the entire ethoxylation train. The reaction is highly exothermic, and adequate cooling must be provided.

Ethoxylation is a heterogeneous phase reaction, and the provision of adequate mass transfer is of prime importance to ensure high productivity and good/consistent product quality.

The autoclaves originally designed used multi-impellers to provide the contact of the liquid phase and the ethylene oxide–vapour phase (stirred autoclaves). Heating (to initiate the reaction) and cooling (to remove the reaction heat) were provided by circulating the reaction mixture in coils in the interior of the autoclaves.

It was later shown that a more intimate contact of the liquid reaction phase and the EO vapours can be achieved by spaying the liquid in finely divided form into an adsorption zone of the autoclave, which is continuously supplied with EO at a rate to maintain a constant pressure in the autoclave (318). The heat of reaction is removed by circulating the reaction mixture through an external heat exchanger and then recycling back to the absorption zone of the reactor. This design uses the Venturi effect to further enhance the contact between the liquid and the vapour phase Venturi-loop reactor. While in the stirred and Venturi-loop reactors, the gas phase is dispersed into the liquid; in the more recent spray tower loop reactor, it is the liquid that is dispersed in the gaseous phase (319).

The most recent reactors proposed are of the hybrid spray tower/Venturi-loop type (320).

8.12.1 ALKYLPHENOLS AND ALKYL NAPHTHOLS

Alkylphenols include octyl-, nonyl-, and dodecyl phenols, of which nonyl phenols are the most widely used.

Alkylphenols were first synthesised in 1940 (321) by Friedel-Crafts alkylation of octene (diisobutene), nonene (propylene trimer), and dodecene (propylene tetramer). In the 1930s, before their use for the synthesis of alkylphenols, these oligomeric olefines were produced essentially as octane modifiers (gasoline polymer) (322).

The production of nonyl phenols has expanded, but since 1985, consolidation and shutting down of smaller, less efficient, and environmentally deficient plants has been the trend in favour of larger, continuous, single-product plants with low-cost operations.

8.12.2 FATTY ALCOHOLS

The production of fatty alcohols relied initially on the ester cleavage of certain natural waxes, for example, spermaceti and sperm oil, which consist mainly of esters of higher fatty alcohols with fatty acids (323). After saponification, the alcohols are separated by distillation with superheated steam or under vacuum or by extraction processes. I.G. Industriefarben in A.G. Ludwigshafen operated a continuous saponification unit of spermaceti (324). This was initially the sole source of fatty alcohols. The volume available was necessarily limited and directed to uses other than the production of surfactants. It should be noted, however, that in some instances, the waxes were directly sulphonated. In this case, the ester linkage was partially split, and the resulting product was a mix of the sulphated alcohols, fatty acids, and esters, which were sometimes neutralised and used without separation.

In the early 1930s, the catalytic hydrogenation of coconut and palm kernel oil was progressed in parallel by Deutsche Hydrier Werke in Germany and E.I. Du Pont in the US. This made larger volumes of fatty alcohols available for sulphation, and fatty alcohol sulphates were the first synthetic detergents used in mass consumer detergent formulations. At that time, Procter & Gamble and Deutsche Hydrier Werke pooled their US interests to form the Hyalsol Corporation and produced several patents for synthesising fatty alcohol sulphates.

The introduction by Procter & Gamble in 1933 into the US market with the washing powder Dreft, based on a fatty alcohol sulphate, was so successful that to keep up with demand, the company had to manufacture its own fatty alcohol using the sodium reduction process for this purpose. The technology was confirmed for the alcohol unit commissioned in 1947 at Ivorydale (Cincinnati, OH). Coconut oil and methyl amyl alcohol were the feedstock. Toluene was used as the medium to melt sodium and as a solvent for the reaction and could be recovered in good yield along with other process aids (325).

Fatty alcohols can be of natural or synthetic origin.

8.12.2.1 Natural Fatty Alcohols

The historic production processes for fatty alcohols were

- The reduction of fatty acids.
- The hydrogenation of fatty acid triglycerides.
- The reduction of fatty acid monoesters.

Today, the reduction of fatty acid monoesters is the route most extensively used.

8.12.2.1.1 Reduction of Fatty Acids

The reduction of fatty acids posed a number of problems, essentially due to (a) the esterification of the freshly formed fatty alcohol and the still unreacted fatty acids and (b) the full hydrogenation of the fatty acid to hydrocarbon (326, 327, 328, 329, 330, 331, 332, 333, 334, 335, 336, 337, 338). The industrial processes eventually adopted operated at 250–300 °C, 150–250 atm with copper, copper chromite, zinc–copper alloys, and copper chromite/iron oxides and are described in several patents (339, 340, 341).

The particular problem posed by oleic acid, that is, the possible reduction of the double bond, was resolved by the use of zinc–vanadium or cadmium–vanadium catalysts at 250–300 °C and 50 atm (342) or by "poisoning" the hydrogenation catalysts, thereby eliminating their effectiveness to hydrogenate the double bond (343).

8.12.2.1.2 Hydrogenation of Fatty Acid Mono Methyl Esters

Already in 1903, Bouveault and Blanc (344, 345, 346, 347) had developed a reduction route for fatty acid esters of lower alcohols. This was based on the generation of native hydrogen through the action of metallic sodium on lower aliphatic alcohols. The modification of the original laboratory process (348) allowed the production of semi-industrial volumes of saturated and unsaturated fatty alcohols.

Despite the industrial hazards associated with the use of sodium and the handling and recovery of low-flash solvents and alcohols, the Bouveault–Blanc method enjoyed a surprising acceptance. In some plants, it was more practicable than catalytic hydrogenation (see the next section on the hydrogenation of fatty acid triglycerides). It also had the advantage of the reactions involved only affecting the carbonyl moiety, leaving undisturbed the double bonds in the hydrophobic chain.

The early process for the catalytic hydrogenation of fatty acid esters with low-molecular-weight alcohols is described in the patent of Adkins, Folkers, and Connor (349).

8.12.2.1.3 Hydrogenation of Fatty Acid Triglycerides

Several processes for the high-pressure hydrogenation of fatty acids are applicable to triglycerides. As early as 1928, the Deutsche Hydrierwerke AG was producing fatty alcohols on an industrial scale using the process patented by Schraut (350). The company supplied the US as well, until in 1933, Du Pont started producing following its patented process (351).

The sodium reduction described in the previous chapter was improved by using C4–C6 secondary alcohols and a solvent, leading to lower operating pressures and temperatures (352) as well as a quantitative utilisation of the sodium (353, 354, 355).

8.12.2.2 Synthetic Fatty Alcohols

The history of synthetic alcohols begins in the 1920s when Franz Fisher and Hans Tropsch discovered the liquefaction of coal to produce branched hydrocarbons (coal-to-liquid [CTL] process).

Today, the main routes to synthetic fatty alcohols are as follows:

- Reduction of fatty aldehydes obtained by the hydroformylation of α-olefins (SHOP oxo alcohols process)
- Hydroformylation of Fisher-Tropsch olefins from the CTL or gas-to-liquid (GTL) processes
- Hydrolysis of aluminium alkyls (Ziegler process)

8.12.2.2.1 The SHOP Process

The Shell higher olefin process (SHOP) produces linear alpha-olefins via ethylene oligomerisation and olefin metathesis invented and exploited by Royal Dutch Shell. It was discovered by chemists at Shell Development Emeryville (CA, United States) in 1968. The research was driven by ecological considerations because of the slow biodegradation of the branched fatty alcohols used widely in detergents, which caused foaming of the surface water, with linear fatty alcohols. At the same time in the US, new gas–oil crackers were commissioned, and the ethylene supply was outpacing demand.

At more than a million tons of capacity, SHOP is one of the standards of industrial olefin metathesis chemistry. SHOP is remarkable in both its chemistry and process engineering. It was developed for the conversion of ethylene to C10–C14 olefins (for conversion to linear primary detergent alcohols by hydroformylation) but can be modified to obtain linear olefins of just about any desired carbon range.

1. *Ethylene oligomerisation*
    In this step, a nickel phosphine catalyst $[(C_6H_5)_2PC_2H_4COO]_4Ni$ is used to oligomerise ethylene given a broad distribution of linear α-olefins. The temperature is in the range of 80–120 °C, and the pressure is between 7–14 Mpa. The mixture is fractionated, and the target olefins (C10–C14) are collected. The light (C4–C8) and heavy (>C14) fractions (50–60% of the mass balance) are transferred to an isomerisation unit.

2. *Olefin isomerisation*

The mixture of light and heavy linear alpha-olefins is passed over an olefin isomerisation catalyst (Na/K on Al_2O_3 or MgO) to give a statistical mixture of linear internal olefins. In this step, for example, 1-octene is converted into 4-octene, and 1-eicocene is converted into 10-eicocene. These are carried forward to a metathesis unit.

3. *Metathesis*

The mixed olefins are subjected to metathesis with an Al_2O_3-supported molybdenum/vanadium/rhodium catalyst. This gives a broad mixture of linear internal olefins, of which 10–15% are in the desired C10–C14 range. For example, 4-octene and 10-eicosene are converted into 2-tetradecene. These are collected from the stream, and the undesired fractions are recycled back to be isomerised.

4. *Ethenolysis*

The internal olefins can also be reacted with excess ethylene with rhenium (VII) oxide catalyst supported on alumina in an ethenolysis reaction which causes the internal double bond to break up and form a mixture of α-olefins with odd and even carbon chain lengths of the desired MW.

5. *Conversion of α-olefins to fatty alcohols*

The C10–C14 α-olefins recovered from the preceding processes are converted to fatty alcohols by the classical process of hydroformylation to aldehyde followed by hydrogenation. In the SHOP process, alcohols are obtained directly due to the greater hydrogenating activity of the catalyst, and the aldehyde hydrogenation step is unnecessary. It is reported that the hydroformylation catalyst was originally $HCo(CO)_3.P(n-C_4H_9)_3$. The trialkyl phosphine increases the linearity of the aldehyde intermediate as well as its conversion into fatty alcohols but causes back-hydrogenation of the olefin feedstock to paraffins.

In 2001, Shell commercialised a variation of this process, using a different catalyst that enabled producing a moderately methyl-branched alcohol with C16–C17 chain length. It is possible that one major detergent manufacturer used this alcohol (or derivatives) in some of its formulations (at least the rumor circulated across the surfactants industry).

It is not confirmed if Shell has implemented a catalyst change in recent years for the mid-cut alcohols and moved to the rhodium/triphenyl phosphine $[HRhCO(PC_6H_5)_3]$ that helps reduce operating pressure and temperature.

8.12.2.2.2   The CTL/GTL Olefin Process

The process is the oxo-carbonylation of olefins produced by the Fischer–Tropsch synthesis of olefins.

The Fischer–Tropsch olefin process is a collection of chemical reactions that converts a mixture of carbon monoxide and hydrogen into liquid hydrocarbons. Sasol's first plant was a 1955 CTL unit (Sasol 1) using Iron catalysts, a carbon copy of the

Ruhrchemie plant in Oberhausen (DE). In 2005 the plant was converted to a natural gas (NG) low-temperature Fischer–Tropsch plant.

Sasol has three technologies:

- Cobalt low-temperature Fischer–Tropsch (Co-LTFT)
- Iron low-temperature Fischer–Tropsch (Fe-LTFT)
- Iron high-temperature Fischer–Tropsch (Fe-HTFT)

These technologies produce fundamentally different types of hydrocarbons and, as a consequence, have the potential to produce different chemical processes. The Fe-HTFT gives the highest content in olefins (55–60% of a mixture of olefins linear and internal branched), followed by n-paraffins (approximately 20%), oxygenated derivatives (5–15% of alcohols, ketones, and carboxylic acids), and aromatics (5–10%).

8.12.2.2.3 The Ziegler Processes

Two processes to produce synthetic fatty alcohols are based on the work of Prof. Dr. Ziegler using organic aluminium compounds: the ALFOL® process, originally developed by Conoco, and the Ethyl Corporation's EPAL® process. Fatty alcohols synthesised by these processes are structurally identical to natural fatty alcohols and are thus ideal substitutes for natural products.

Conoco (now Sasol) started the first ALFOL plant in the US in 1962. The plant is still operating and capitalises on cheap ethylene feedstock that offsets the issues arising from its age. In 1964, Condea Chemie (now Sasol) built a similar plant in Brunsbüttel (Germany) which was de-bottlenecked in 2008. Additional ALFOL alcohol plants were built in Russia and China. Ethyl Corporation (today BP/Amoco) developed its own process (EPAL process) and started operations in 1964.

The process Involves five steps: hydrogenation, ethylation, growth, oxidation, and hydrolysis.

Steps 1–4 are carried out in a hydrocarbon solvent (heptane). Two thirds of the triethylaluminum produced in the ethylation reaction (2) are recycled into the hydrogenation stage (1), and one third enters the growth reaction (3). The insertion of the ethylene molecule into the aluminium–carbon bonds occurs as a statistical process and leads to a Poisson distribution of chain lengths, ranging from C_2 to beyond C_{26}. An optimum yield of the C_{12}–C_{14} alcohols, which are important in the surfactant sector, requires the addition of about 4 molecules of ethylene per aluminium–carbon bond. A small percentage of olefins are formed as by-products.

Due to the varying reactivity of partially oxidised trialkylaluminum compounds, oxidation is carried out stepwise by passing the compounds through carefully dried air. Cooling is necessary, especially at the start of the reaction. Alkanes and oxygen-containing compounds are formed as by-products.

Prior to hydrolysis, the solvent is removed by distillation. Hydrolysis with water gives high-purity hydrated alumina, which has many industrial applications, for example, in catalytic processes and ceramics. In the 1960s, hydrolysis was carried out using hot sulphuric acid at Conoco and Ethyl Corporation plants. Conoco changed to neutral hydrolysis, but the sulphuric acid method is still used in the EPAL process and leads to high-purity aluminium sulphate as a co-product. The crude alcohols are finally fractionated into marketable blends and single cuts.

1) *Hydrogenation*

$$2 \; Al(CH_2CH_3)_3 + Al + 1.5 \; H_2 \; \rightarrow \; 3 \; HAl(CH_2CH_3)_2$$

2) *Ethylation*

$$3 \; HAl(CH_2CH_3)_2 + 3 \; CH_2CH_2 \; \rightarrow \; 3 \; Al(CH_2CH_3)_3$$

3) *Growth Reaction*

$$3 \; Al(CH_2CH_3)_3 + (x+y+z) \; CH_2CH_2 \; \rightarrow \; Al \begin{array}{l} (CH_2CH_2)_x\text{-}H \\ / \\ - (CH_2CH_2)_y\text{-}H \\ \backslash \\ (CH_2CH_2)_z\text{-}H \end{array}$$

4) *Oxidation*

$$\begin{array}{l} R_1 \\ / \\ Al - R_2 + 1.5 \; O_2 \; \rightarrow \\ \backslash \\ R_3 \end{array} \quad \begin{array}{l} OR_1 \\ / \\ Al - OR_2 \\ \backslash \\ OR_3 \end{array}$$

5) *Hydrolysis*

$$\begin{array}{l} OR_1 \\ / \\ Al - OR_2 \; + \; 2H_2O \; \rightarrow \; AlO(OH) + R_1OH + R_2OH + R_3OH \\ \backslash \\ OR_3 \end{array}$$

FIGURE.8.3   Scheme of the ALFOL process.

Source: Author contribution.

EPAL PROCESS

Many attempts have been made to achieve a narrower distribution of chain lengths in the growth reaction. The only process that has been used at an industrial scale is the EPAL process developed by the Ethyl Corporation (today BP/Amoco). The reaction steps resemble those of the ALFOL process, but the growth reaction is not carried as far. The product of the growth reaction is subjected to trans-alkylation (290 °C, 3.5 mPa) with C_4–C_{10} olefins. The chain lengths of the resulting trialkylaluminium compounds are predominantly C_4–C_{10}. Excess olefins are removed in a stripping column and then fractionated. The trialkylaluminum compound is subjected to a second growth reaction and then trans-alkylated (200 °C, 35 kPa) with C_{12}–C_{18} olefins. Again, the olefins are separated in a stripper and fractionated. At this stage, the trialkylaluminium compound consists largely of alkyl chains with 12–18 carbon atoms.

8.12.2.2.4   Fatty Alcohol Consumption

The consumption of fatty alcohols from renewable resources is on the rising, because of environmental, sustainability, the and attitude of the personal care industry that increasingly demands sustainable products and a general perception across mass consumers that "natural" is safer and has a better overall image than the "synthetics". While most regions of the world have now moved to natural, in the US, there is still a large consumption of synthetics, driven by their large, installed capacities and low ethylene prices.

8.12.3   EO

EO was first prepared in 1859 by Wurtz by dehydrohalogenation of 2-chloroethanol (356). The initial industrial processes were based on this chemistry: acetylene from the hydrolysis of calcium carbide was reduced to ethylene which was further reacted with hypochloric acid, and the resulting 2-chloroethanol was converted to EO by the action of hot sodium hydroxide water solutions (357, 358). The direct oxidation of ethylene dates from 1931 (359), and today, most of the EO is produced by silver-catalysed oxidation in air-based or oxygen-based processes (Figure 8.4).

For a long time, Wurtz's 1859 synthesis remained the only method for preparing ethylene oxide, despite numerous attempts, including by Wurtz himself, to produce ethylene oxide directly from ethylene (360). The original process or a variant based on the dehydration of ethanol to ethylene followed by silver oxidation continued to be used for the production of limited volumes when ethylene was not readily available.

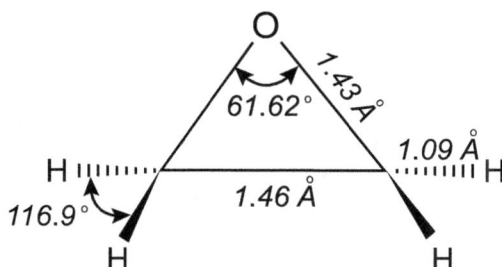

FIGURE 8.4   Structure of ethylene oxide.

Source: Author adaptation from M.J. Schick (ed.), *Nonionic Surfactants*, Marcel Dekker, New York (1968).

Commercial production of EO dates back to 1914 when BASF built the first factory which used the chlorohydrin process, schematically as shown:

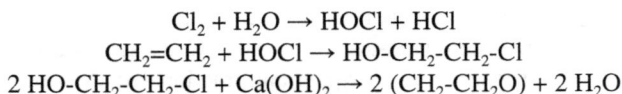

$$Cl_2 + H_2O \rightarrow HOCl + HCl$$
$$CH_2=CH_2 + HOCl \rightarrow HO-CH_2-CH_2-Cl$$
$$2\ HO-CH_2-CH_2-Cl + Ca(OH)_2 \rightarrow 2\ (CH_2-CH_2O) + 2\ H_2O$$

Today, the quasi-totality of the EO is produced by silver-catalysed oxidation in air-based or oxygen-based processes. Union Carbide (now a division of Dow Chemical Company) was the first company to develop industrially the direct oxidation process (361). It was further improved in 1958 by Shell Oil Co. by replacing air with oxygen and using elevated temperatures of 200–300 °C and pressure (1–3 mPa) (362).

In recent years, the synthetic route from ethanol has been resumed, to produce sustainable ethylene oxide. Ethanol from the fermentation of molasses is reduced to ethylene which is further oxidised via the classical silver-catalysation process. Scientific Design offers key-on-door plants (363), and Croda International plc has built a production unit in its plant at Wilmington (Delaware) for an investment of 170 million USD, that has been recently commissioned.

The direct oxidation of ethanol to EO is under investigation at the present (364).

The industrial production of ethylene oxide involves the following steps:

- Ethylene oxidation: This is generally done by passing ethylene through bundles of tubes, each 6–15 metres long, 20–50 mm in diameter, packed with a catalyst in the form of spheres or rings 3–10 mm in diameter. The prevailing operating conditions are 200–300 °C and a pressure 1–3 MPa. Temperature control is vital. With the ageing of the catalyst, its selectivity decreases, and it produces more exothermic side-products of CO_2.
- EO scrubbing: The gaseous stream from the main reactor, containing EO (1–2%) and CO_2 (around 5%), is cooled and then passed to the EO scrubber. Water is used as the scrubbing media which scrubs away the majority of EO, along with some amounts of CO_2, N_2, CH_2CH_2, and CH_4 and aldehydes (introduced by the recycle stream). Also, a small proportion of the gas leaving the EO scrubber (0.1–0.2%) is removed continuously (combusted) to prevent the build-up of inert compounds (N_2, Ar, C_2H_6), which are introduced as impurities with the reactants.
- EO de-sorbtion: The aqueous stream from the scrubbing is sent to the EO de-sorber. EO is obtained as the overhead product, whereas the bottom product (known as the 'glycol bleed') that contains mono-ethylene glycol, di-ethylene glycol, and other poly-ethylene glycols is continuously bled off.
- Stripping and distillation: The EO stream is stripped of its low-boiling components and then distilled in order to separate it into water and EO.
- CO_2 scrubbing and CO_2 de-scrubbing: The recycle stream from the EO scrubber is compressed and fed to a solution of potassium carbonate in the CO_2 scrubber. The CO_2 reacts with the potassium carbonate:

$$K_2CO_3 + CO_2 + H_2O \rightarrow 2\ KHCO_3$$

The solution of potassium bicarbonate is then sent to the CO_2 de-scrubber, where CO_2 is de-scrubbed by stepwise (usually two steps) flashing. The first step is done to remove the hydrocarbon gases, while the second step strips off CO_2.

8.12.4 PO

PO was also first synthesised in Wurtz's laboratory, by converting dibromopropane to propylene glycol, subsequently converting it to 1-chloro-2-propanol followed by dehydrochlorination with diluted potassium hydroxide.

In the first industrial process (chlorohydrin process), the chlorohydrin is prepared by the reaction of propylene and chlorine in excess water followed by dehydrohalogenation. Because of the large volume of effluent produced in the process, the direct oxidation of propylene with hydroperoxides in the presence of a catalyst (the oxirane process) was later developed (365, 366, 367):

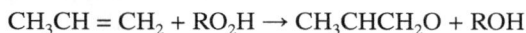

$$CH_3CH = CH_2 + RO_2H \rightarrow CH_3CHCH_2O + ROH$$

The process is practised with three hydroperoxides (368):

- tert-Butyl hydroperoxide is derived from the oxygenation of isobutane, which produces t-butyl alcohol. This co-product can be dehydrated to isobutene and converted to methyl tert-butyl ether.
- Ethylbenzene hydroperoxide is derived from oxygenation of ethylbenzene, which gives 1-phenylethanol as a by-product. This co-product can be dehydrated to give styrene.
- Cumene hydroperoxide is derived from the oxygenation of cumene which gives cumyl alcohol as a by-product. Via dehydration and hydrogenation, this co-product can be recycled back to cumene. This technology was commercialised by Sumitomo Chemical (369).

In March 2009, BASF and Dow Chemical started up their new HPPO plant in Antwerp (Belgium), where propylene is directly oxidised with hydrogen peroxide.

Also, Evonik and ThyssenKrupp Industrial Solutions have jointly developed a process for the production of PO from propylene and hydrogen peroxide (370).

Between 60–70% of all PO produced is polypropoxylated and end EO tipped to give the building blocks for producing polyurethanes (371).

About 20% of PO is hydrolysed into propylene glycol, and other major products are polypropylene glycols, propylene glycol ethers, and propylene carbonate.

8.12.5 FATTY ACIDS

These were dealt in Chapter 2, "Soap-making process and raw materials".

8.12.6 FATTY AMINES

These are dealt in Section 7.2 "Cationic surfactant production process and raw materials".

BIBLIOGRAPHY

(257)   US. Pat. 1,970,578
(258)   US. Pat. 2,231,477
(259)   US Pat. 2,174,761
(260)   TOXNET, Toxicology Data Network
(261)   EPA, "Nonyl Phenol and Nonyl Phenol Ethoxylates Action Plan [RIN 2070
        – ZA09]"
(262)   Margel, M., Toxipedia, Nov. 01, 2011
(263)   Griffin, W.C., "Classification of Surface-Active Agents by 'HLB" *Journal of the
        Society of Cosmetic Chemists* 1(5), pages 311–326 (1949), archived from the original
        on 2014–08–12, Retrieved 2013–05–25
(264)   Griffin, W.C., "Calculation of HLB Values of Non-Ionic Surfactants" *Journal of the
        Society of Cosmetic Chemists* 5(4), pages 249–256 (1954), archived from the original
        on 2014–08–12, Retrieved 2013–05–25
(265)   Davies, J.T., "A Quantitative Kinetic Theory of Emulsion Type, I. Physical Chemistry
        of the Emulsifying Agent Gas/Liquid and Liquid/Liquid Interface" in Shulman, J.H.,
        ed. *Proceedings of the International Congress of Surface Activity*, Academic Press,
        New York, pages 426–438 (1958)
(266)   Xiawen, Guo, Zhongming, Rong, Xugen, Ying, "Calculation of Hydrophile-Lipophile
        Balance for Polyethoxylated Surfactants by Group Contribution Method" *Journal of
        Colloid and Interfacial Science* 298 (1), pages 441–450 (2006)
(267)   Mollet, M., et al., "An Efficient Method to Determine the Hydrophile-Lipophile
        Balance of Surfactants Using the Phase Inversion Temperature Deviation of CiEj/n-
        octane/water Emulsions" *International Journal of Cosmetic Science* 41(2), pages
        99–108 (2019)
(268)   US Pat. 2,322,822
(269)   US Pat. 2,374,931
(270)   Jap. Patent 4015246
(271)   US Pats. 2,374,931; 2,380,166
(272)   Fischer, E. "The Compounds of Sugars with Alcohols and Ketones" *Berichte* 23,
        1145–1895, Collective volume I, pages 740–741 (1923)
(273)   US 2,049,758 A
(274)   US 2,390,507
(275)   US 4,889,925
(276)   Hill, K.H., von Rybinski, W., *Alkyl Polyglyglycosides, Technology, Properties and
        Applications* G. Stoll, ed., VCH Verslagsgesellschaft mbH, D-69451, Weinheim,
        page 221 (1997)
(277)   Wieland, H. Chem. Ber., 54B, page 2368
(278)   Oswald, A.A., Guertin, D.L., J. Org. Chem., 28, (1963) page 651
(279)   US Pat. 2,169,976
(280)   US Pat. 2,999,068 (1961); South African Pat. 61–1798 (1961); US Pat. 3,001,945(1961);
        French Pat. 1,378,770 (1964)
(281)   Lake, D.B. and Hoh, G.L.K., J. Am. Oil Chemists Soc. 40, page 628 (1963)
(282)   Lindner, H., Tenside 1, page 112 (1964)
(283)   Jungermann, E., and Ginn, M.E., Soap Chem Specialties, 40 (9), page 59 (1964)
(284)   Belgian Pat. 626,346
(285)   British Pat. 380,431 (1932)
(286)   French Pat. 727,202 (1932)
(287)   US Pat. 1,970,578 (1934)
(288)   German Pats. 605,973 (1934); 667,744 (1938)

(289) French 770,804 (1934)

(290) Brit. Pat. 476,571 (1937)

(291) Belgian Pat. 445,808 (1942)

(292) US Pats. 2,290,880 (1942); US Pat 2,269,314 (1954); US Pat 2,091,105 (1937)

(293) German 844, 449 (1952)

(294) Swiss Pat. 263,840 (1949)

(295) Malkemus, J.D., J. Am. Oil Chemists Soc., 33, page 571 (1956)

(296) Scholz, D.H.J., Sthuler, D.H., et al., Hoechst AG (1989)

(297) Behler, A.S., Rahts W.N.S., et al., Henkel KGaA (1994)

(298) Leach, J.C.P., Lin et al., Vista Chemical Company (1993)

(299) Hama, I., Okamoto T., et al. J. Am. Oil Chem. Soc. 72, pages 781–794 (1995)

(300) Behler, A., Syldat A., *5th World Surfactants Congress*, Florence, May 29–June 2 (2000), page 405

(301) Herczuch, W., Krasnodebsky, Z., et al. *Instytut Ciezkej Syntezy Organicznej*, Blachownia (Poland) (1995)

(302) Szymanowski, J. Przemyl Chemivznj 84, page 567 (2005)

(303) U.S. Pat 5,220,046 (1993)

(304) Proprietary Mg/Al Catalysts, Brochure from Lion Specialty Chemicals Co. Ltd.

(305) Wurtz, C.A., "Mémoires sur l'Oxide d'Ethylene et les Alcohols Polyéthyléniques" *Annales de chimie et de physique* 69, pages 317–354 (1863)

(306) Staudinger, H., Schweizer, O., "Ueber Hochpolymere Verbindungen, 20 Mittel", "Ueber die Poly-Aethylene Oxide" *Berichte der Deutschen Chemischen Gesellschaft B* 62, pages 2395–2405 (1929)

(307) Flory, P.J., "Molecular Size Distribution in Ethylene Oxide Polymerisation" *Journal of the American Chemical Society* 62, pages 1561–1565 (1940)

(308) US Pat. 3,036,130 (1962)

(309) US Pat. 2,677,700 (1954)

(310) Matlock, P.L., Brown, W.L., Clinton, N.A., *Synthetic Lubricants and High-Performance Functional Fluids, Revised and Expanded*, 2nd ed., Rudnick, L.R., Shubkin, R.L., eds., CRC Press, Boca Raton, FL (1999)

(311) Greaves, H., Zang-Hoozemans, E., Kelidj, N., van Voorst, R., Meertens, R., "Performance Properties of Oil Soluble Synthetic Polyalkylene Glycols" *Lubrication Science* 24, pages 251–262 (2012)

(312) US 2,129,709; US 2,205,021 (18.06.1938)

(313) US 2,619,466 (1952)

(314) Schulze W.A., Lyon, G.P., Ind. Eng. Chem. 40, page 2308 (1948)

(315) Iwai, S.J., Nippon Oil Technol. Soc., 2 (6), 19 (1949)

(316) US 2,347, 564 (1944)

(317) Sachat, N., Greenwald, H.L., *Nonionic Surfactants*, Schick, M., ed., Chapter 2, Marcel Dekker Inc., New York, pages 8–12 (1967)

(318) US Pat. 2,586,767 (1952)

(319) Di Serio, M., Tesser, R., Santacesaria, E., "Comparison of Different Reactor Types Used in the Manufacture of Ethoxylated/Propoxylated Products" *Industrial & Engineering Chemistry Research* 44, pages 9482–9489 (2005)

(320) Santacesaria, E., Tesser, R., Di Serio, M., "Polyethoxylation and Polypropoxylation Reactions: Kinetics, Mass Transfer and Industrial Reactor Design" *Chinese Journal of Chemical Engineering* 26(6), pages 1235–1251 (2018)

(321) Fiege, H., Voges, H.W., Hamamoto, T., Sumio, U., Iwata, T., Miki, H., Fujita, Y., Buysch, H.J., Garbe, D., Paulus, W., "Phenol Derivatives" in *Ullmann's Encyclopedia of Industrial Chemistry*, Wiley-VCH, Weinheim (2002)

(322) Eglof, G., Ind. Eng. Chem. 28 (12), pages 1461–1467 (1936)

(323) French Pat. 776,044

(324) Stüpel, H., *Syntetische Wash-und Reinigungsmitteln*, Konradin-Verlag, Robert Kohlhammer, Stuttgart, 2 Auflage, page 89 (1957)

(325) Stüpel, H., *Syntetische Wash-und Reinigungsmitteln*, Konradin-Verlag, Robert Kohlhammer, Stuttgart, 2 Auflage, page 88 (1957)

(326) Schraub, Chemiker-Ztg, 55, 3, 16 (1931)

(327) Seifensieder Ztg 58, 61 (1931)

(328) Angew. Chem. 46, 459 (1931)

(329) Ger. Pats 607,792; 626,979; 629,244; 636,681

(330) French Pats.701,200; 703,844

(331) Schraub, Schenk und Stickdorn, Ber. Dtsch. Chem. Ges. 64, 1314 (1931)

(332) Adkins und Folkers, J. Amer. Chem. Soc. 53, 1095 (1931); 54, 1145 (1932)

(333) Adkins and Connor, J. Amer. Chem. Soc. 53, 1091 (1931)

(334) US pat. 2,091,800

(335) Schmidt, Ber. Dtsch. Chem. Ges. 64, 2051 (1931)

(336) Ger. Pat. 573,604

(337) Normann, Angew. Chem. 44, 471, 714, 922

(338) French Pat. 708,286

(339) US Pat. 1,839,974

(340) Ger. Pats. 670,832; 594,481; 617,542

(341) French Pat. 718,394

(342) US Pat. 2,374,379

(343) French Pat. 799,097

(344) Bouveault, Compt. Rend. 136, 1676 (1903); 137, 60.328 (1903); Bull. Soc. Chim. (3) 31, 666.1206 (1904)

(345) Ger. Pat. 164,294 (1903)

(346) US Pat. 868,252 (1907)

(347) Blanc, A., Perfumer 17, 57

(348) US Pat. 2,019,022

(349) US Pat. 2,091,800

(350) Ger. Pats. 56,471; 56,488

(351) US Pats. 1,746,783; 1,964,000; 1,984,884; 2, 0794, 414; 2,105,540; 2,109,844

(352) US Pats. 1,971,742/43; 2,096,036

(353) Hansley, J., Amer. Chem. Soc., 57, page 2303 (1935)

(354) Hansley, J., Ind. Eng. Chem. 39, 55 (1947)

(355) US Pats. 2,070,597; 2,096,036; 2,158.071

(356) Stanley, H.M., *Ethylene and its Industrial Derivatives*, S.A. Miller ed., Ernest Benn, London, pages 16, 17, 527, 528 (1969)

(357) Stüpel, H. *Synthetische Wasch-und Reinigungsmittel*, 2nd ed., Konradin Verlag, Stuttgart, page 153 (1975)

(358) Cawse, J.N., Henry, J.P., Swatzlander, M.H., Wadia, P.H., *Kirk-Othmer Encyclopaedias of Chemical Technology*, 3rd ed., vol. 9, M. Grayson ed., Wiley-Interscience, New York, page 432 (1980)

(359) Lenher, S., J. Am. Chem. Soc., 53, 3737, 3752 (1931)

(360) "Ethylene Oxide" *Kirk-Othmer Encyclopedia of Chemical Technology. Elastomers, Synthetic to Expert Systems*, 9, 4th ed., John Wiley & Sons, New York, pages 450–466 (1994)

(361) Bloch, H.P., Godse, A., *Compressors and Modern Process Applications*, John Wiley & Sons, New York, 2006, pages. 295–296, ISBN 978-0-471-72792-7

(362) Weissermel, K., Arpe, H.-J., *Industrial Organic Chemistry*, 4th ed., Wiley-VCH, Weinheim, 2003, pages 145–148. ISBN, 978-3-527-30578-0

(363) *Renewable EO and Glycol*, Scientific Design Company Inc. Brochure

(364) Lippits, M.J., Nieuwenhuis, B.E., "Direct Conversion of Ethanol into Ethylene Oxide on Copper and Silver Nanoparticles" *Catalysis Today* 154, pages 127–132 (2010)

(365) Kuhn, W., *Hydrocarbon Process* 59(10), page 123 (1979)

(366) Scheldon, R.A., Van Doorn, J.A., *J. Catal.* 34, page 242 (1974); Scheldon, R.A., Van Doorn, J.A. *J. Catal.* 31, page 427 (1973)

(367) Simmrock, K.H., "Die Herstellverfaren fur Propyleneoxide und ihre elektroche-mische Alternativen" *Chemie Ingenieur Technik*, 48, page 1085 (1976)

(368) Nijhuis, T.A., Makkee, M., Moulijn, J.A., Weckhuysen, B.M., "The Production of Propene Oxide: Catalytic Processes and Recent Developments" *Industrial & Engineering Chemistry Research* 45, pages 3447–3459 (2006)

(369) "Summary of Sumitomo Process from Nexant Reports". Retrieved 2007–09–18

(370) Thyssen-Krupp Industrial Solutions Brochure

(371) Adam, N., et al., "Polyurethanes" in *Ullmann's Encyclopedia of Industrial Chemistry*, W71ey-VCH, Weinheim (2005)

9 Other Surfactants

9.1 SILICONE-BASED SURFACTANTS

Silicone surfactants consist of a permethylated siloxane moiety, coupled with one of molar polar groups, which can be anionic, nonionic cationic or amphoteric. Although the early patents on the structure and applications were granted at the end of the 1950s, they found prior use in the production of polyurethane foams.

Polyurethane foams were first produced in the laboratories of I.G. Farbenindustrie A.G. By Dr. Otto Bayer and his co-workers in 1937 and commercialised in the early 1950s (372). The economic production of polyurethane foams with specific performance properties requires the use of silicon-based surfactants. These are active at the liquid–air, liquid–liquid, and polymer–polymer interfaces within the foams and can reduce the surface tension to 20 mN/m. This cannot be achieved with hydrocarbon-based surfactants that are typically in the range of 26–28 mN/M.

The surface active character of siloxane surfactants is due to the high -CH_3 density attached to the -O-Si-O-Si-, which provides also a very flexible backbone. The surface energy of a methyl-saturated surface is about 20 mN/m, which is also the lowest surface tension achievable with siloxane surfactants. Hydrocarbon surfactants contain mostly methylene -CH_2- groups which have a higher surface energy.

In 1958, the preparation of organosilicone polyether copolymers was patented by the Union Carbide Corporation (373). Fritz Hostettler of Union Carbide introduced the use of silicone polyether polyols copolymers in polyurethane foams (374).

The first silicone surfactants were prepared by transesterification of an alkoxymethyl siloxane with hydroxy-terminated polyoxyalkylenes (375) but these hydrolise rapidly at non-neutral pH to yield silanol and alcohol. More hydrolytically stable surfactants are produced by the hydrosilylation of methyl siloxanes containing Si–H groups with vinyl functional polyoxyalkylenes (376, 377, 378):

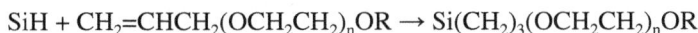

$$SiH + CH_2=CHCH_2(OCH_2CH_2)_nOR \rightarrow Si(CH_2)_3(OCH_2CH_2)_nOR$$

Hydrosilylation of methyl siloxanes.

Another possibility for attaching a polar group to a siloxane backbone is to replace the Si–H group with a reactive organic group and to which the polar group is further attached (see previously mentioned references).

Th Goldschmidt AG (now Evonik) pioneered the silicone surfactants which are now available from several producers.

Beside polyurethane foams, silicone surfactants have found applications as demulsifiers in oil production and in fuels, textile and fabrics lubricants and softeners, hair conditioners, and in paints and coatings as wetting agents, flow promoters, and antifoams and deaerators.

DOI: 10.1201/9781003403869-11

9.2 FLUOROSURFACTANTS

Fluorosurfactants are short carbon chains partially or totally fluorinated and terminated with a hydrophilic group that can be anionic, nonionic, amphoteric, or cationic.

They have distinctive features and deliver effects not matched by the hydrocarbon-based surfactants. This is because

- The low surface energy of the hydrophobic moiety enables repelling both oil and water. Hence, fluorosurfactants exhibit both oil and water repellency when absorbed on substrates such as paper and textiles.
- The strength of the C–F bond (116 kcal/mole in CF4) imparts an exceptional stability in extreme conditions of pH, temperature, and chemical environment. A further contribution to stability comes from the small atomic radius of the covalently bound fluorine atoms (0.72 Å). Fluorine atoms can then shield carbon atoms practically without any steric stress.
- Fluorosurfactants can lower the surface tension of aqueous systems to below 20 mN/m and are effective at exceptionally low concentrations. Only 10 ppm of a fluorosurfactant may be needed to lower the surface tension of water to 40 mN/m. They also exhibit surface activity in organic media.

Fluorosurfactants behave as wetting agents for low-energy surfaces. On higher-energy surfaces, the fluorosurfactants can absorb with the fluorocarbon chain oriented towards the solution and behave as de-wetters. This dual behaviour is also noted with respect to foaming. Some fluorosurfactants (e.g., from the class of the amphoterics) are excellent foaming agents. Anionic and nonionic molecules are low-foaming, and in some instances, the anionics can function as antifoam agents.

Commercially important pathways to fluorosurfactants are electrochemical fluorination and the telomerisation and oligomerisation of tetrafluoroethylene.

Electrochemical fluorination is described in publications and patents of the late 1940s and early 1950s (379, 380, 381, 382).

Commercial telomerisation of tetrafluoroethylene with pentafluoroethyl iodide was developed by the du Pont Company in the early 1960s (383, 384). However, the production of fluorinated alcohols

$$H(CF_2CF_2)_nCH_2OH$$

by the telomerisation of methanol with tetrafluoroethylene in the presence of a peroxy or azo catalyst was already described in du Pont patents dating back to the early 1950s (385, 386).

The anionic polymerisation and oligomerisation of tetrafluoroethylene is described in Brit. Patent 1,302,350 (1973). The anionic polymerisation catalysed by a fluoride salt produces highly branched oligomers. The main products are tetramers, pentamers, and hexamers. These were commercialised under the trade name Monflor.

The presence of fluorinated alkyl chains in ionic or nonionic surfactants has led to products that exhibit pronounced lower surface tensions and good wetting properties. For example, N-ethyl-N-polyoxyethylene ethanol perfluoro octane sulphonamide

$$C_8F_{17}SO_2N(C_2H_5)(C_2H_4O)_nH$$

in an aqueous solution at 0.001% can reduce the surface tension to 21.6 mN/m (387).

An extensive review of the fluorinated surfactants, synthesis, properties, and applications can be found in *Fluorinated Surfactants*, Kissa, E. ed., Surfactants Science Series, vol. 50, Marcel Dekker, New York (1994).

9.3 ACETYLENIC SURFACTANTS

Acetylenic surfactants are variations of the basic structures of 3,6 dimethyl 4-octyne 3,6 diol or 2, 4, 7, 9 tetramethyl 5-decyn 4,7 diol (TMDD).

The OH groups can be ethoxylated, thus enabling the fine-tuning of the hydrophilic–lipophilic balance of the molecule.

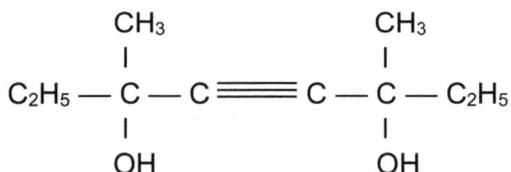

3,6-dimethy1-4-octyne-3,6-diol

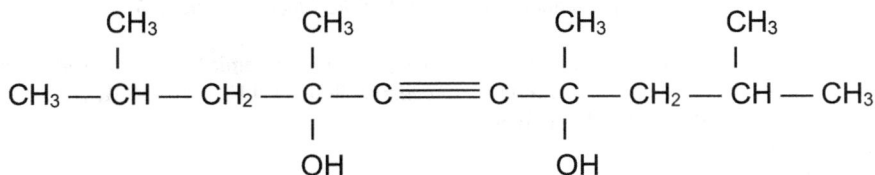

2,4,7,9-tetramethy1-5-decyn-4,7-diol

FIGURE 9.1 Acetylenic diols.

Source: Author contribution.

In these molecules, the triple bond and the oxygens of the two symmetrical hydroxyl or polyoxyethylene groups create a domain of high electronic density, which imparts hydrophilic properties. The highly branched alkyl chains are the hydrophobic moieties. The six-methyl groups in TMDD reduce the attraction between adjacent molecules and prevent the formation of micelles in aqueous solutions. As a consequence, all of the surfactant present in the solution is available for interfacial effects. Also, the symmetrical configuration with short hydrocarbon chains enables the molecules to orient horizontally at the air–fluid interface and form expanded films that can withstand surface pressures up to 30 mN/m without collapsing.

Acetylenic surfactants are unique in that they are excellent wetting agents (due to their mobility, high surface area per molecule – 50–250 square Å – and surface orientation) and, at the same time, mild foamers or defoamers (because acetylenic surfactants can exert high surface pressure, they can displace other foaming surfactants from the surface).

The synthesis of acetylenic surfactants was already disclosed in 1935 (388) and further perfected in the early 1940s (389) and 1960s (390). Surfynol 104 (a trademark of Air Reduction Co. Inc.) was the most prominent acetylenic glycol used for its surface active properties. Its ethoxylation is described in British 893,431 (to Air Reduction Co.) and led to a range of nonionics with increased water solubility.

BIBLIOGRAPHY

(372)   Herrington, R., Hock, K., ed., *Flexible Polyurethane Foams*, Dow Plastics, Midland, MI (1991)
(373)   US Pat. 2,834,748 (1958)
(374)   Ger. Pat 1,091,324 (1960)
(375)   Snow, S.A., Fenton, W.N., Owen, M.J., *Langmuir* 6, page 385 (1990)
(376)   Plumb, J.B., Atherton, J.H., *Block Copolymers*, Allport, D.C., Janes, W.H., ed., Applied Science Publisher, London, page 305 (1973)
(377)   Noll, W., *The Chemistry and Technology of Silicones*, Academic Press, New York (1968)
(378)   Clarson, S.J., Semlyen, J.A., ed., *Siloxane Polymers*, PTR Prentice Hall, New York (1993)
(379)   Simons, J.H., et al., "Production of Fluorocarbons: I. The Generalized Procedure and its use with Nitrogen Compounds" *Journal of the Electrochemical Society* 95, page 47 (1949)
(380)   *Fluorine Chemistry*, Simons, J.H., ed., vol. I, page 414, Academic Press, New York, 1950
(381)   Simons, J.H., Brice, T.J., *Fluorine Chemistry*, Simons, J.H., ed., vol. II, page 340, Academic Press, New York (1954)
(382)   US Pat. 2,519,983 (1950)
(383)   US Pat. 3,145,222 (1961)
(384)   US Pat. 3,226,449 (1965)
(385)   US Pat. 2,559,628 (1951)
(386)   US Pat. 2,559,629 (1951)
(387)   Burnette, L.W., *Nonionic Surfactants*, Schick, M.J., ed., Surfactants Science Series, vol. 1, Marcel Dekker Inc., New York, page 418 (1967)
(388)   US Pat. 2,106,180 (1935)
(389)   US Pat. 2,250,445 (1941)
(390)   US Pat. 3,108,140 (1963)

Part III

Washing and Cleaning Habits

10 Washing Linen and Clothes

For centuries in antiquity, in the Middle Ages and in some countries even to date, linen and cloth washing was just the cumbersome process of beating clothes, smashing them on rocks, or beating them with sticks. In Europe, this was still the case until the not-too-distant past.

Greeks and Romans did not wash their clothes at home and used to send their garments to specialised workers called by the Romans Fullo or Nacca (fullers, fullones) (391).

Although the Romans knew about soap, as documented by the description of the naturalist Pliny (*Historia Naturalis* 28.li.191) and by the later references of Galen, Serenus Sammonicus, and Theodorus Priscianus, they did not use it for washing clothes, preferring instead to use other kinds of alkali. Of these, the most common one was the urine of animals or humans, the latter was collected in vessels placed at the corners of the streets to be filled by the passengers (392) and diluted with water by people doing the cleaning (called "fullones") in the washing bath (393). Suetonius (394) reports that the emperor Vespasian imposed a "urinae vectigal", which is supposed to have been a tax paid by the fullones. The alkaline material collected was mixed with water (395), and "creta fullonia", fuller's earth, was also used (396) probably in part to provide a scouring action, in part to absorb the greasy dirt.

The fullones cleaned garments that had been already worn as well as new clothes from the loom to be scoured and smoothed. The way in which the cleaning was done as described by Pliny and other writers and is clearly explained in paintings found on the walls of a fullonica at Pompei. The fullones first trod upon and, with their feet, stamped on the clothes immersed in the alkaline washing solution in tubs or vats, a rhythmic movement described by Seneca (397) as "saltus fullonicus" (the jumping of the fullones). It is probable that the clothes were thoroughly rinsed thereafter and hung out to dry in the streets before the doors of the fullonica (398). When dry, the clothes were brushed and carded to raise the nap, hung on a vessel of basketwork (viminea cavea), and exposed to a steam of sulphur vapours to destroy unwanted colours and stains (399). A fine white earth (called Cimolian by Pliny) was often rubbed into the clothes to enhance whiteness (400).

An establishment or workshop of fullers was called a Fullonica (401), Fullonicum (402), or Fullonium (403). Like other principal businesses in Rome, the fullones formed a "collegium", and their activity was considered so important to deserve the attention of the Censors C. Flaminius and L. Aemilius who, in 220 BCE, prescribed the mode in which clothes were to be washed (404).

What we know about cloth washing in the Middle Ages is desperately scarce. There are scattered references to the use of plants rich in saponins and ashes, but probably only the dominant nobility practised clothes washing. Quite frankly, it is difficult to imagine that a lot of clothes washing occurred among a peasantry barely having clothes. The reference to soap in the *Capitulare de Villis* does not specify its intended use, but it is my view that it may have been used to treat textiles, especially wool.

Through the centuries, there was an excruciatingly slow process of moving from mechanic treatment to the use of washing aids (ashes, saponins, possibly a little soap). As an example, as late as the early 20th century, it was common practice in large households in the Italian countryside to do one major clothes washing once a year. Shirts, underwear, and linen were boiled with ashes in large copper cauldrons, rinsed, dried, and ironed. The washed garments and linen covered the expected use for a full year. Except for occasional washing jobs and for more delicate garments, soap was not used.

The advent of formulated detergent first (Henkel's "Persil"), synthetic detergents, and washing machines completely changed washing habits.

BIBLIOGRAPHY

(391)   Theophr. Char. 10, Athen xi, 528d; Pollux vii., 39,40,41; Martial, xiv, page 51
(392)   Martial vi., 93; Macrob. Saturn, ii, page 12
(393)   Pliny, Historia Naturalis xxxviii 18.26, Athen, xi, page 484
(394)   Suetonius, Vesp, page 23
(395)   Pliny, Historia Naturalis, xxxi, page 46
(396)   Pliny Historia Naturalis xvii, page 4 and xxxv, page 57
(397)   Seneca, Ep. 15
(398)   Dig. 43, tit 10s1 paragraph 4
(399)   Apul. Met. ix, page 208; Pliny Historia Naturalis xxxv, page 50,57; Pollux, vii., page 41
(400)   Theophr. Char., page 10; Plaut., Aulul. iv, page 9.6; Pliny, Historia Naturalis xxxv., page 57
(401)   Dig. 39, tit.3s3
(402)   Dig.7, tit.1s13, paragraph 8
(403)   Amm. Marc., xiv
(404)   Pliny, Historia Naturalis xxxv, page 57

11 Body Washing and Personal Hygiene

In all times and cultures personal hygiene, cleanliness and body care were a direct reflection of social class, although it is remarkable that for considerable periods of the European and American history living unwashed was common among the masses, clergy, nobles, saints, and monarchs alike.

Archaeological evidence suggests the existence of 5000-year-old bathing facilities in Gaza. Today, the ruins at the site of Mohenjo-Daro in the region of Sindh (Pakistan) and dated to 2500 BCE are possibly the oldest standing public bathhouse. As we have seen soap-like material found in clay jars of Babylonian origin has been dated to about 2800 BCE, although it is most likely that the intended use of the stuff was as an auxiliary in textile dyeing.

Before the time of Abraham, in Middle Eastern desert climes, custom dictated that hosts offer washing water to guests to clean their feet.

One of the first known bathtubs comes from Minoan Crete; it was found in the palace at Knossos and is dated about 1700 BCE. The palace plumbing system had terra-cotta pipes jointed and cemented together that were tapered at one end to give the water a shooting action to prevent the build-up of clogging sediments.

We do not have evidence that the ancient Egyptians developed similar structures, but they liked hygiene, which was evident in their use of fresh linen, body ointments, skin conditioners, and perfumes.

Ritual bathing was part of ancient (and still is of modern) Jewish culture. Ritual cleansing baths (*mikvot*) from the classical period have been found in archaeological digs at multiple sites, including Masada. The distinctive nature of *mikveh* structures causes them to be regarded as archaeological markers of Jewish communities at classical and medieval sites. A mikveh dating from around 1150 has been uncovered by archaeologists in Bristol, England, and another in Cologne, Germany, dates from around 1170 (405).

As with other ancient people, the Greeks apparently did not use soap but anointed their bodies with oil and ashes, scrubbed with blocks of pumice or sand, and scraped themselves clean with a curved metal instrument called a "strigil". Immersion in water and anointment with olive oil completed the ablutions.

Mentions of sweat baths go back to Roman accounts of the Scythians' habit of taking sweat baths: "These tents were made of thick felt, with all cracks carefully sealed up. Inside was placed a bowl full of red-hot stones, onto which cannabis seeds were thrown". According to Herodotus, the Scythians would howl with delight as they breathed in the fumes. Sitting in these tents was clearly one of their favourite pastimes (406).

To the Romans, access to and the supply of clean water in the urban communities as an aspect of sound administration, personal hygiene and cleanliness were

DOI: 10.1201/9781003403869-14

almost a prerequisite for participating the public and economic life. Rome, as the capital of the empire, may well have been over-represented in both the extent of the facilities available and their recording and reporting, but taken in due perspective, it is a fair representation of a general pattern. In the 4th century CE, the city had 11 large and magnificent public bathhouses, more than 1350 public fountains and cisterns, and many hundreds of private baths. Served by 13 aqueducts, Rome's per capita daily water consumption is estimated at a remarkable 1600 litres.

By the peak of their popularity, the thermae included hot and cold rooms and medium-temperature lounging rooms, with a variety of extra services such as food, wine, exercise, and/or personal training being offered. At different points in the history of Rome, baths were gender-segregated by place or time, while at other times, the bathing was mixed (407).

Baths and thermae were not only places to care for the body: they offered at the same time the opportunity to do business or to forge political alliances under a cover of discretion and to develop and maintain a social network. As observed, the Roman baths were a daily social activity, in the same way that modern teenagers frequent the local swimming pool and adults the exercise club.

The disintegration of the Roman Empire caused the degradation of the social, political, and economic systems and their supporting fabrics: roads and communications, buildings, culture and the arts, aqueducts, public and private buildings, and, of course, thermae. With the rapidly declining welfare and in times during which populations were decimated by invasions, epidemics, wars, and famine, it is hardly surprising that personal care and hygiene must have become a low, and probably last, priority. In Europe, the diffusion of Christianity and, within it asceticism and monasticism, contributed to making bathing a mundane, sinful habit. In response to the perceived debauchery and the association to pagan practices of Roman baths, the early Christian church frequently discouraged cleanliness. Saint Jerome (340–420 CE) made the biting comment that "He who has once bathed in Christ has no need of a second bath". To those that are well, and especially to the young, Saint Benedict in the 6th century commanded that "bathing shall seldom be permitted". The fathers of the early Christian Church condemned bathing as "unspiritual" and equated bodily cleanliness with the luxuries, materialism, paganism and what has been called "the monstrous sensualities of Rome".

Ostracism for bathing was not absolute; for example, Gregory the Great, the first monk to become pope, allowed Sunday baths and even commended them, so long as they did not become a "time-wasting luxury". Nevertheless, the comments of other great personalities like Saint Jerome quoted earlier must have undoubtedly left some marks. A lack of care for the body was considered a Christian virtue and living unwashed one of the many facets of sanctity. Saint Francis of Assisi considered an unwashed body a stinking badge of piety.

Roman-type baths were continued and/or re-established in Islamic countries through the medieval and Renaissance periods, and bathing was endorsed by Islamic writers. The hammam, referred to in modern times as the "Turkish Bath", was a major feature of Islamic culture and preserved the Roman traditions of cleaning the body first and then soaking and socialising. Due to the Islamic religious requirements for frequent washing (when water was unavailable, dust or sand could be used for ritual ablutions), baths and washing equipment remained popular. Some historians believe that the habit of the baths returned to Western Europe from the Middle East with the Crusaders, but documentary evidence suggests that the resurgence of public baths in Western Europe may have been more a function of political and economic stability (408).

Japanese baths and bathing habits have been, for centuries, a distinct feature of that nation. Western writers claim that the soaking baths of Japan originated from the extensive use of Japanese hot springs (409):

> Situated between two volcanic belts, Japan offers countless natural thermal baths. The tradition of public bathing dates back at least to the mid of the sixth century and to the dawn of Buddhism, which taught that such hygiene "not only purified the body of sin but also brought luck".

According to an article in the *Economist* (410):

> Chinese historians commented in the third century on the cleanliness of the Japanese. So did European travelers when Japan first began to open to them at around the turn of the 16th century. Until the emergence of public baths in the 17th century, nearly all the baths for the common folk were provided by Buddhist temples.

A considerable part of this chapter is devoted to personal hygiene, bathing, and cleanliness in the Middle Ages in Europe, partly because this is relatively well documented and partly in an attempt to answer the question, Is it really true that Europe during the Middle Ages went 1000 years without a bath?

There are partisans supporting this and the opposite view with remarkable passion. In my opinion, there is an inherent bias in the way the debate is progressed: either party has tried to prove its thesis and refute that of the opponents by referring to specific segments of the society, nobility as opposed to the populace, urban as opposed to rural settlers. Of course, focusing on one quartile alone produces a much-distorted picture that can range from perfect cleanliness to absolute filth. Trying to draw a picture across the board highlights the following:

- Bathing as a social ritual was relatively popular and practised. There seems to be an emerging consensus that bathing was more a matter of social mores than for hygiene.
- Church regulations on bathing were designed to combat excessive indulgence in the habit and the immoral practices that ensued from promiscuity.
- Deprived of the sophisticated Roman infrastructure, well-off medieval and Renaissance people appear to have bathed less often but with the same social enjoyment.

The upper classes, as represented by the medieval nobility, appear to have routinely washed their hands before and after meals. Etiquette guides of the age (which, of course, were accessible and directed to the minority of the better-off) insisted that teeth, face, and hands be cleaned each morning. Shallow basins and water jugs for washing hair were found in most manor houses, as was the occasional communal tub.

Georges Duby, in an article in *A History of Private Life* (411) suggests:

> Among the dominant classes at least, cleanliness was much prized. In the eleventh and twelfth centuries, the Cluniac monasteries and houses of the lay nobility continued to set aside space for baths. . . . No formal dinner (that is, no dinner given in the great hall with a large crowd of guests) could begin until ewers had been passed around to the guest for their pre-prandial ablutions. Water flowed abundantly in the literature of amusement – over the body of the knight-errant, who was always rubbed down, combed, and groomed by his host's daughters whenever he stopped for the night, and over the nude bodies of fairies in fountains and steam-baths. A hot bath was an obligatory prelude to the amorous games described in the fabliaux. Washing one's own body and the bodies of others seems to have been a function specifically ascribed to women, mistresses of water both at home and in the wilderness.

A saying in France from 13th century shows how bathing was considered one of the pleasures of existence: "Venari, ludere, lavari, bibere; Hoc est vivere!" (To hunt, to play, to wash, to drink, – this is to live!) (412).

It has been rightly pointed out that care should be taken not to attribute to the 12th–15th centuries the disgusting lack of cleanliness of the 16th and subsequent centuries which, in most countries in Europe and European colonies, continued up to the 19th and early 20th centuries.

However, in the Middle Ages bathing and grooming were regarded with suspicion by moralists, however, because they unveiled the attractions of the body. Bathing was said to be a prelude to sin and in the penitential of Burchard of Worms, we find a full catalog of the sins that ensued when men and women bathed together. Lambert of Ardres, the historian of the Counts of Guînes, describes the young wife of the ancestor of his hero swimming before the eyes of her household in a pond below the castle, but he is careful to indicate that she was wearing a modest white gown. Public baths were suspect because they were too public; it was better to wash one's body in the privacy of one's own home. Scrupulous, highly restrictive precautions were taken in monasteries. At Cluny, the custom required the monks to take a full bath twice a year, at the holidays of renewal, Christmas and Easter, but they were exhorted not to uncover their pudenda.

Two aspects emerge from the documentary literature of the time:

• The bathing of mixed genders
• The extent to which bathing facilities were available in urban areas

There are many illustrations (reported for example in Frank Crisp's *Medieval Gardens*", New York, Hacker Art Books [1966]), depicting groups or individuals bathing in fountains or spring pools, or in the series of men and women bathing by

Durer from the end of the 15th century (413). One interesting illustration (*The Baths at Louèche*, by Hans Bock the Elder, 1597) depicts mixed bathers in a sort of pool, some of them in amorous attitudes while others appear to be minding their own business; for instance, one composed young woman is reading a book. It is hard to tell whether these are artistic fantasies or interpretations of a common practice, but it sure that by the 15th and 16th centuries, bathhouses in Western Europe had mixed clientele, and that by the end of that period, the public baths had widely gained the ill repute that has come to our times.

Russian bathing habits have several peculiarities. Steam baths in wooden bath-houses in Russia are mentioned by the apostle Andreas in the Russian Primary Chronicle of 1113 (414):

> They warm them to extreme heat, then undress, and, after anointing themselves with tallow, take young reeds and lash their bodies. They lash themselves so violently that they barely escape alive. Then they drench themselves with cold water and thus are revived. They think nothing of doing this every day and actually inflict such torture upon themselves voluntarily.

Traditions in Russia of bathing before weddings are confirmed in the *Domostroi*, and the tradition of giving birth in the banya, or bathhouse, appears to originate before 1600 as well.

According to Levin (415):

> The usual location chosen for the delivery, at least in Northern Russia, was the bathhouse. Archbishop Ilia of Novgorod in the twelfth century formulated specific instructions on how to purify it, citing the precedent of Bishop Nifont. The bath-house was warm, clean, and private. It could be placed off limits of the delivery and cleansing without disrupting village routines. Furthermore, the bathhouse had a religious significance in Finno-Ungaric paganism. It served as a center for gathering, worship, religious dance, and personal re-purification. The custom of giving birth in a bath-house was ingrained to the point that women in the seventeenth century who gave birth out of wedlock and killed the newborn still went to the bath-house for the delivery.

Apparently, in Russia bathing in a tub was part of the bride's pre-wedding duties, as (416):

> On the eve of the appointed day, the bride's mother, girlfriends, and female rela-tives arranged a ritual bath for her. After the bride had washed, the bath water was saved; it was supposed to have magical powers transferred from her body to excite love in her future husband. Although Orthodox clergy tried to stamp out the cus-tom of collecting the "wedding water" the custom continued into the seventeenth century.

While we do not know whether this bath would involve immersing the whole body, it's clearly one that involves standing water rather than, or in addition to, steam or dry heat.

Washing at other times is attested in the 16th/17th-century text of the *Domostroi*. The author mentions washing upon arising and before praying at the start of a new task. Ritual bathing and washing also appear in the *Domostroi* text. The 17th-century wedding ritual sections mention the groom's post-nuptial visits to the bathhouse and the washing of the bride inside the house. The exchange of ceremonial decorated "towels" also appears in the *Domostroi*'s descriptions of nuptial arrangements.

It is reported that in Russia bathing could involve interactions between people of the opposite sex despite church rules. According to Pushkareva:

> Contemporary observers reported that, in the tsar's household, the tsar and his retainers might meet in the bath-house, which provided both bathing facilities and a sauna. However, it is unlikely that the women of the tsar's family, and much less women of the lower classes, followed the same custom. Women did visit the bathhouse, but it was usually on holidays or on Saturday evenings. The tsaritsa and her daughters had their own section of the palace bathhouse. The Stoglav Church Council of 1551 prohibited men and women, monks and nuns, from bathing together,' proclaiming those who did so as 'without shame.' But the common people did not observe this prohibition, and men and women bathed naked together. In the winter, they ran out of the bathhouse naked to roll in the snow to cool off, without regard for curious onlookers. It took more than a century before Russian bathhouses were divided into separate men's and women's baths.

While mixed bathing was discouraged by the Catholic Church, records exist that baths were used as social affairs, with banquets and wedding feasts being joined with the baths. Certainly, the depictions of couples using the baths suggest that it was a social, as well as sexual, activity. However, there are indications that several ordinary public bathhouses were often segregated by gender or that different times or days were restricted for each gender.

Around the 13th and 14th centuries, public bathhouses were established in European cities. Paris, for example, numbered 26 steam baths and bathhouses in 1292 (417). They were commonplace enough to attract the attention of authorities and prompt the issue of regulations (see the following discussion). It was not uncommon to offer sessions in a steam bath as a tip to artisans, domestic servants, or day-laborers. A crier patrolled the streets of 13th-century Paris to summon people to the heated steam baths and bathhouses. Typical services offered were a steam bath, with in addition, according to price, a bath in a tub, wine, a meal, or a bed.

In Bohemia, Poland, and Germany, public baths, mainly steam baths, seem to have been a fixture. In 1385, when Jadwiga of Poland was apparently concerned about the appearance of her prospective bridegroom, Jagiello of Lithuania, "she was only placated after a favorite young knight of hers, Zawisza of Olesnica, had been sent to inspect Iogaila (Jagiello) in his bath-house and reported back favorably on the details of the barbarian's body" (418). Always, according to the same author, in the 1400s, "there were no less than twelve public baths in Krakow".

German vapour baths were known in the 12th century, as Hildegarde of Bingen suggests herbs in mixtures to pour over the head in the sauna, splash on the sauna rocks, apply to the body and/or drink in the bath, and bathe in (419). A face-wash recipe in her *Physica* reads:

> But one whose face has hard and rough skin, made harsh from the wind, should cook barley in water and having strained that water through a cloth, should bathe his face gently with the moderately warm water. The skin will become soft and smooth and will have a beautiful color. If a person's head has an ailment, it should be washed frequently in this water, and it will be healed.

The regulations governing the guild of bathhouse keepers in Paris around 1270 (420) prescribe the following:

1. Whoever wishes to be a bathhouse-keeper in the city of Paris may freely do so, provided he works according to the usage and customs of the trade, made by agreement of the commune, as follows.
2. Be it known that no man or woman may cry or have cried their baths until it is day, because of the dangers which can threaten those who rise at the cry to go to the baths.
3. No man or woman of the aforesaid trade may maintain in their houses or baths either prostitutes of the day or night, or lepers, or vagabonds, or other infamous people of the night.
4. No man or woman may heat up their baths on Sunday, or on a feast day which the commune of the city keeps. And every person should pay, for a steam-bath, two deniers; and if he bathes, he should pay four deniers. And because at sometimes wood and coal are more expensive than at others, if anyone suffers, a suitable price shall be set by the provost of Paris, through the discussion of the good people of the aforesaid trade, according to the situation of the times. The male and female bathhouse-keepers have sworn and promised before us to uphold these things firmly and consistently, and not to go against them.
5. Anyone who infringes any of the above regulations of the aforesaid trade must make amends with ten Parisian sous, of which six go to the king, and the other four go to the masters who oversee the trade, for their pains.
6. The aforesaid trade shall have three good men of the trade, elected by us unanimously or by a majority, who shall swear before the provost of Paris or his representative that they will oversee the trade well and truly, and that they will make known to the provost of Paris or his representative all the infringements that they know of or discover, and the provost shall remove and change them as often as he wishes.

Washing the hands before eating is emphasised in the manners manuals of the 15th and 16th centuries. It was also recommended to wash the hands and face in the morning upon arising and to wash out the mouth with cold water.

Ceremonial washing at the beginning of meals was ritualised, but at the end of the meals, it was necessary and usual; at formal tables, basins, pitchers, and towels were brought to the attendants, although, in some cases, a separate handwashing area may have been set up for diners to visit.

Washing before praying or before going to bed was not as common. In fact, according to *A Drizzle of Honey*, washing before praying or at night before bed could mark a Spanish *converse*, during the time of the Inquisition, as a recusant Jew or Muslim as these were Jewish and Islamic requirements (421).

Babies were bathed in tubs: many depictions of the Birth of the Virgin show the infant Mary being bathed or wrapped after the bath (422). Smaller basins were used for partial washing, such as those being presented to Saint Anne for washing after childbirth in the Birth of Virgin paintings and the Italian "childbirth" commemorative trays (423).

In depictions of hospitals and sickrooms, nurses are sometimes shown washing patients in a sort of sponge bath or bed bath, with cloth and basin.

We have a very vivid and accurate description of the cleaning habits among the lower classes, especially the shepherd and the peasantry, from the accounts that have come to our days of the detailed records of the Inquisition Register of Jacques Fournier, bishop of Pamiers in Ariège (southern France) from 1318–1325. Jacques Fournier (later to become Pope Benedict XII) conducted a rigorous Inquisition in his diocese to eradicate the last fringes of the Cathar heresy. A methodical personality, he ensured that the depositions made to the Inquisition courts were accurately recorded. In revealing their position on official Catholicism, the peasants examined by Fournier (many from Montaillou, the last village that actively supported the Cathar heresy) gave an extraordinary, detailed picture of their everyday life.

To quote Emmanuel Le Roy Ladurie, professor of history of modern civilisation at the College de France, Paris, who made an accurate exert of the voluminous Inquisition records (3 volumes) in "Montaillou, Cathars and Catholics in a French Village, 1294–1324" (424):

> In Montaillou, people did not shave, or even wash, often. They did not go bathing or swimming. On the other hand, there was a good deal of delousing.
>
> If delousing was common, ablutions were so summary as to be almost non-existent. Crossing water was very dangerous and people did not bathe or swim. They may be lingering near the baths at Aix-les-Thermes [an urban centre near Montaillou, n.d.a.], but that was only to sell sheeps or visit the prostitutes. The springs themselves, which were of the simplest description, were mainly reserved for lepers and people with ringworm . . .
>
> In Montaillou the toilet, when it existed, took no notice of the anal or genital areas, but was restricted to those parts of the body which blessed, handled, or swallowed food – i.e. the hands, face, and mouth. To give water to someone's hands [in italic in the text n.d,a.] was considered a sign of politeness.
>
> So, the inhabitants of Montaillou took off their linen at night [these seem to have consisted, among others, of "shirts" and "underpants", n.d.a.]. They even changed it sometimes! At considerable intervals, Pierre Mauray had a change of shirt brought to him in the pastures by his brother Arnaud. The fact struck him as sufficiently remarkable to be included in his deposition [to the Inquisition authorities, n.d.a.]. But we cannot say how often people changed, or washed, their clothes. . . .
>
> Clothes were certainly considered precious. Jean Maury walked for several days over mountainous country to bring Bélibaste . . . the patched-up garments of a dead friend.

Note that the cleaning habits (or rather the lack of it) of the peasantry in Montaillou were also common to the lower nobility living in the village. Thus, the location of settling (urban vs rural) was as an important differentiator as the belonging to a social class.

Following the devastating outbreak of the Black Death in England (1348–1350), a link appears to have been made between health and hygiene. In 1388, the English parliament issued the following statute in an effort to improve what appears to be a disastrous situation:

> Item, that so much dung and filth of the garbage and entrails be cast and put into ditches, rivers, and other waters . . . so that the air there is grown greatly corrupt and infected, and many maladies and other intolerable diseases do daily happen. It is accorded and assented, that the proclamation be made as well in the city of London, as in other cities, boroughs, and towns through the realm of England, where it shall be needful that all they who do cast and lay all such annoyances, dung, garbage, entrails, and other ordure, in ditches, rivers, waters, and other places aforesaid, shall cause them utterly to be removed, avoided, and carried away, every one upon pain to lose and forfeit to our Lord the King the sum of 20 pounds.

By the 16th century, health concerns provided support for the idea that bathing was a dangerous practice to be avoided. It undoubtedly contributed to the appalling situation of personal hygiene in the 17th and 18th centuries. The plagues and other widespread diseases of the Middle Ages inspired a belief that water was a carrier of infection. Most people went to great lengths to avoid contact with water. Their objective was to maintain an appearance of cleanliness by keeping clean what was visible. Masking odours with perfumes and powders was the substitute. Although six bathrooms were installed in the Palace of King Louis XIV (other sources say two), he is said to have bathed only twice in his life and to have been terribly ill after each occasion.

Colonial America's leaders deemed bathing impure, since it promoted nudity, which could only lead to promiscuity. Laws in Pennsylvania and Virginia either banned or limited bathing. For a time in Philadelphia, anyone who bathed more than once a month faced jail. The first bathtub was brought in by Benjamin Franklin in the 1780s. The first three-piece bathroom, with a toilet, tub, and washbasin, emerged in Philadelphia in 1810.

For English people of that time, bathing was very problematic. There was no running water, streams were cold and polluted, heating fuel was expensive, soap was hard to get or heavily taxed, and there were no facilities for personal hygiene. Cleanliness was not part of the folk culture. At the end of the 18th century, running water was available only in a few aristocratic homes, and cleanliness was considered a luxury.

The association of hygiene and health began to be accepted, supported by the growing claims of the medical establishment that bathing was beneficial and the positive statements of intellectuals like Balzac. The value of bathing received powerful reinforcement in the 1870s when the discoveries of Louis Pasteur confirmed the existence of a hidden bacterial world.

At the turn of the 19th century, a change in bathing practices began to take place. Running water began to work its way into everyday life. City-wide plumbing systems brought water directly into homes, first only into basements to be carried up in buckets. However, technology soon took advantage of water pressure, and water distribution to upper floors became possible. Running water, effective sewer systems, increased standards of living, and more disposable income for non-essential goods brought a step-change in personal hygiene practices. More important, it raised the consciousness of the benefits of cleaning across larger and larger layers of the population to reach the present standards.

In many countries, after centuries, a private bathroom in every home has become a reality.

BIBLIOGRAPHY

(405)   Aldous, T., "Bristol's Judaica Project" *History Today* 47(7), pages 27, 28 (July 1997)

(406)   James, P., Thorpe, N., *Ancient Inventions*, Ballantine, New York, page 342 (1994)

(407)   Garrett, F., *Bathing in Public in the Roman World*, University of Michigan Press, Ann Arbor (1999)

(408)   Haise, A.J., *A Short History of Bathing Before 1601* (2007)

(409)   Von Furstenberg, D., *The Bath*, Random House ed, New York, page 91 (1993)

(410)   "Very Clean People, the Japanese" *The Economist*, 344 (8028), page 66 (August 2, 1997)

(411)   Duby, G., "Solitude: Eleventh to Thirteenth Century" in *A History of Private Life: Volume 2: Revelations of the Medieval World*, Georges Duby, ed., Belknap Press, Cambridge, MA, pages 509–534 (1998)

(412)   Vigarello, G., Birrell, J., *Concepts of Cleanliness: Changing Attitudes in France Since the Middle Ages*, Cambridge University, New York (1988)

(413)   Vigarello, G., Birrell, J., *Concepts of Cleanliness: Changing Attitudes in France Since the Middle Ages*, Cambridge University, New York, page 23 (1988)

(414)   Allen, N., "May the Steam be with You" *Russian Life* 47(1), pages 22–29 (January–February 2004)

(415)   Levin, E., "Childbirth in Pre-Petrine Russia" in *Russia's Women: Accomodation, Resistance, Transformation*, University of California Press, Berkeley (1991)

(416)   Pushkareva, N., *Women in Russian History from the Tenth to the Twentieth Century*, trans. and ed. Eve Levin, M.E. Sharpe, Armonk, NY (1997)

(417)   Riolan, *Curieuses Recherches*, page 219

(418)   Zamoyski, A., *The Polish Way: A Thousand-Year History of the Poles and Their Culture*, Hippocrene, New York, page 43 (1987)

(419)   Hildegarde of Bingen, *Hildegard von Bingen's Physica: The Complete English Translation of Her Classic Work on Health and Healing*, trans. from the Latin by Patricia Throop, Healing Arts, Rochester, VT (1998)

(420)   Taken from Etienne de Boileu, *Livre des métiers*, translated in *Women's Lives in Medieval Europe*, Pearson Education Limited, London, ISBN 978-1-138-85567-0

(421)   Glitlitz, D.M., Davidson, L., *The Lives and Recipes of Spain's Secret Jews*, St. Martin Press, New York, September 2000

(422)   Thornton, P., *The Italian Renaissance Interior, 1400–1600*, H.N. Abrams, New York (1991)

(423)   Musacchio, J., *The Art and Ritual of Childbirth in Renaissance Italy*, Yale University Press, New Haven (1999)

(424)   Penguin Books, 1980, pages 141,142, English translation of the 1978 edition by Editions Gallimard, Paris, of *Montaillou, Village Occitan de 1294 á 1324*

Index

Symbols and Numbers

β-alkylamino-propionic acid, 32
1-chloro-2-hydroxyalkan, 44
1-chloro-2-propanol, 91
1-hydroxyethyl-2-octadecylimidazoline
        oxide, 77
2-chloroethanol, 89
2-ethylhexyl diphenyl phosphate, 54
2-octadecyl-1-(3-sulphopropyloxyethyl)
        imidazoline-1-oxide, 77
2-tetradecene, 86
3-dimethylamino propyl amine (DMAPA), 32
4-octene, 86
10-eicosene, 86

A

Abelard of Bath, 8
acetylenic surfactants, 99
acyl chlorides, 48, 56
acyl glutamates, 58–59
acyl glycinates, 59
acyl halides, 56
acyl sarcosinates, 58
acyl taurates, 57–58
Adams, John, 14
ALFOL process, xii, 87–88
Alipal, 38
alkylaryl sulphonates, 40–44
        alkyl benzene sulphonates, 41–43
        alkyl diphenyl oxide (di)sulphonates, 43–44
        alkyl naphthalene sulphonates, 40–41
        naphthalene sulphonate-formaldehyde
                condensates, 41
alkyl (paraffin) sulphonates, 44–45
alkylpolyglucosides (APGs), 75–76, 88
alkylsulphochlorides, 44–45
ammonium peroxide, 77
amphoteric surfactants, 31–35
        production process and raw materials,
                32–35
anionic surfactants, 35–60
        alkylaryl sulphonates, 40–44
        alkyl (paraffin) sulphonates, 44–45
        dicarboxylic (sulphosuccinates) and
                tricarboxylic sulphonated esters, 40
        fatty alcohols sulphation process and raw
                materials, 39–40
        isethionates, 48–49
        methyl esters sulphonates, 46–47

olefin sulphonates, 45–46
phosphate esters, 53
sulphated alkyl, aryl, and alkylaryl ethers,
        38–39
sulphated alkyl esters, 36
sulphated amides, 37–38
sulphated fatty alcohols (alkyl sulphates), 38
sulphated glycerol esters, 36–37
sulphated oils and fatty acids, 35–36
sulphated olefins, 38
sulphonation process and raw materials
        50–51
surfactants based on amino acids and protein
        hydrolysates 56–57
APGs, see alkylpolyglucosides
aquatic toxicity, xii, xiii, 72, 80
ashes
        barilla, 8, 10
        body cleaned with, 4, 105
        clothes-washing in, 103–104
        slurry of, 3
        soaps made with mixtures using, 5, 7, 9,
                13–14
        wood, 14, 20
Avirol AHX, 36

B

bathing
        American Civil War 15
        ancient Rome, xi, 9
        babies, 116
        bathhouses, 106, 110–111
        bathtubs, 105
        bubble baths, 57
        Christianity and 106, 110, 116
        Colonial America, 117
        health and, 117
        impurity of morals associated with, 117
        Middle Ages, 6, 106–108
        normalizing of, 118
        France, during reign of Louis XIV, 13
        Puritan New England, 14
        Russia, 109–110
        sponge bath, 116
        steam baths, 110
        steam baths as social norm, xx
        smegmata, 4
        sweat baths, 105
Bayer Leverkusen, 45
Bayer, Otto, 97

benzene, 40, 50, *see also* alkyl benzene
    sulphonates
    alkylated, 51
    ethylbenzene, 91
benzene giving 9- or 10-phenylstearic acid, 43,
    50–51
bile acids, 27
black char, 52
Black Death, 117
black soap, 10, 13

C

castor acid, 44
castor oil, 35
cationic surfactants, 67
    production process and raw materials, 67
Charlemagne, 6, 9
Charles I, 11
cholic acid, 27
choline phosphatidyl, 27
coal, 20
    burning of, to heat bathwater, 111
    gasification of, xii
coal tar, 50
coal-to-liquid (CTL) process, 85
Colgate & Company 15–16
Colgate-Palmolive-Peet Co. 16, 36–37
    APGs in products by, 76
    Gardol (sodium lauroyl sarcosinate) used
        by, 58
Colgate, William 15
Conoco, 87
copper, 84
copper chromite catalyst, 69

D

David of Antioch, 8
dicarboxylic (sulphosuccinates) and tricarboxylic
    sulphonated esters, 40
Diderot and D'Alembert, 20
diesters, 49, 75
    as poor hydrotropes, 54
    diester of methyl diethanolamine (MDEA), 69
    diester of triethanolamine (TEA), 69
dimethyl amine, 69
dialkyl amine, 67
dimethylamine oxide, 77
dimethyl sulfate, 69–70
Dow Chemical Company, 43–44, 80, 91

E

EO, PO, and EO/PO homo – and copolymers,
    79–80, 89–91
    EO/PO derivatives, 81

EPAL process, xii, 87–88
ester quats, 69–70

F

FAMEs, *see* fatty acid methyl esters
fatty acid esters of polyhydric alcohols and
    alkoxylated derivatives 72
    HLB concept, 73–74
fatty acid methyl esters (FAMEs), 78–79
fatty alcohols, 83–89
    alkoxylation of, 71
    natural, 84–85
    sulphation process and raw materials,
        39–40
    synthetic, 85–89
fatty amines, 31–32, 60, 91
    alkoxylated, 71
    polyoxyethylene fatty amines, 78
    primary, 68
    salts, 67
    secondary, 69
fatty nitriles, 69
fluorosurfactants, 98
Fremy (pupil of Gay-Lussac), 29
Friedel-Crafts synthesis, 42–44, 50–51, 83

G

Galen, 6, 9, 103
Galileo, 20
Gay-Lussac, 29
Gladstone, William 11
Günther and Hetzer, 40

H

Henkel and Cie, 40, 45, 76, 78
    Persil, xi, 23, 104
Humectol CX, 37
Hyalsol Corporation, 83
hydrocarbon chain 53, 79
    olefins of, 38
hydrocarbons
    aliphatic, 46
    aromatic, 36, 50
    branched, 85
    Fischer-Tropsch, 45, 50
    gases, removal of, 91
    hydrogenated, 68, 84
    liquid, xii, 86
    lubricants, 80
    polynuclear aromatic, 40
    radicals, 44
    Sasol technologies, 87
hydrochloric acid, 20, 42, 52
hydrosylilation, 97

I

Igepal 38
Igepal NA, 42–43, 50
IG Industriefarben, 83
isethionates, 48–49
isopropanol, 40, 59
    as solvent, 70
isopropyl halides 40
isopropyl naphthalene sulphonate, 29, 40

J

jodation, 42
jodobenzene, 42

K

Krafft process, 40, 42
Krafft temperature, 59

L

Lambert of Ardres 108
Leblanc process, 20
Leonil, 38
Lever Brothers 12, 16
Louis XIV, 13, 117
Ludwig the Pious 9
Lurie, 42
lye, 4, 7, 8, 10, 14

M

Malherbe, 20
Methrie, de la, 20
methyl esters sulphonates (MESs), 46–48
    tallow, 46
    three stages of manufacture of, 47

N

Naccosol A, 41
Naccosol LAL, 48
Naccosol NR, 42
naturally occurring surface active agents, 25–27
    bile acids, 27
    phospholipids, 27
    saponins, 25–27
natural surface active agents
    French soap, 8
    soap, 3, 23
    soap made from olive oil or tallow, 7
    soap-making process and raw materials, 19
Nekal, 29
Nekal A, 41
Nekal B or B, 41

Nekal NS, 29, 49
nonyl phenol ethoxylates (NPEs), 72
nonionic surfactants, 71
    alkoxylation of fatty alcohols and alkyl
        phenols, 71
    alkylpolyglucosides, 75–76
    amine oxides, 76–78
    EO, PO, and EO/PO homo – and copolymers,
        79–80
    ethoxylated fatty amines, 78
    ethoxylated mercaptans, 80–81
    ethoxylated methyl esters, 78–79
    fatty acid esters of polyhydric alcohols and
        alkoxylated derivatives, 72–74
    nonionic surfactants production process and
        raw materials, 81–92
    polysorbates, 75
    sorbitan esters, 74–75
nonionic surfactants production process and raw
        materials, 81–92
    alkylphenols and alkyl naphthols, 83
    EO, 89–91
    fatty acids 91
    fatty alcohols, 83–89
    fatty amines 91
    natural fatty alcohols, 84–85
    PO, 91
    synthetic fatty alcohols, 85–89
NPEs, *see* nonyl phenol ethoxylates

O

olefin sulphonates, 45–46
oleic acid, 29, 35–37, 84
    benzene giving 9- or 10-phenylstearic acid
        reacting with, 43, 50–51
    bisulfite reacting with, 44
    oleic amide, 58
oleic sorbitan mono esters, 75

P

palmitoyl chloride, 42
PEGs, *see* polyethylene glycols
Pliny, 8
phosphate esters, 53
phospholipids, 27
POEFA, *see* polyoxyethylene fatty amines
Poisson or Poisson-type distribution, 79, 87
polyethylene glycols (PEGs), 79
polyethylene oxide polypropylene oxide
        copolymers (EO-PO-EO), (PO-EO-PO),
        80–81
polyoxyethylene fatty amines (POEFA), 78
polysaccharides, xix, 25
polysorbates, xx, 75
polysulphides, 44

Pompadour, Madame de, 13
Procter & Gamble, 15–16, 76, 83, 84
    Camay, 16
    Chipso, 16
    Ivory soap, 16
Procter, Harley, 16
Procter, William, 15

Q

quaternary ammonium, 67
    compounds, 68, 76
    esters, xxi
    fatty, 31
quaternary ammunium salts (quats), 67

R

ricinoleic acid, 36, 44, 49
rosin, xix, 15, 19, 50

S

saponins, xix, 15, 25–27, 103–104
Sasol, 87
Schiff's base, 32
Scythians, 105
Serenus Sammonicus, 6, 103
Settimius Severus, 6
silicone surfactants, 97
soap
    French, recipe for making, 8
    hard, discovery of and recipe for making 8–9
    history of, 3–7
    later developments in production of, 23
    olive oil or tallow, recipe for making 7
soap-making process and raw materials, 19
    alkali, 19
    history of 4
    oils, fats, and fatty acids, 19
sodium
    crude, 4
    harder soap, as compared to potassium, 20
    metallic, 84
    reduction process, 84, 85
    Wurtz synthesis with jodobenzene and, 42
sodium bromoethane sulphonate, 31
sodium carbonate, 8, 20, 21, 23, 67
sodium cetyl sulphate, 38
sodium chloroacetate, 32
sodium isethionate, 48
sodium laurel sarcosinate, 58
sodium methoxide, 78
sodium perborate, 23
sodium salts, 27

sodium silicate, 23
sodium sulphate, 20, 52
sodium methyl sulphate, 47
sodium sulphite, 44, 48
sodium bisulphite, 50
Solvay process, 20
sorbitan esters, 74–75
Stockhausen, Julius, 35
Stockholm Papyrus, 9
Suetonius, 103
sulphamic acid, 52
sulphated alkyl, aryl, and alkylaryl ethers, 38–39
sulphated alkyl esters, 36
sulphated amides, 37–38
sulphated fatty alcohols (alkyl sulphates), 38
sulphated glycerol esters, 36–37
sulphated oils and fatty acids, 35–36
sulphated olefins, 38
sulphochlorination, 45
sulphonation process and raw materials, 50–51
    alkylation of an aromatic moiety, 50–51
    sulphation and sulphonating agents, 51
sulphuric acid, 20, 29, 35–43
    as catalyst, 50, 76
    direct sulphonation of aliphatic hydrocarbon
        with, 46
    EPAL process, 87
surfactants based on amino acids and protein
        hydrolysates, 56–60
    acyl glutamates, 58–59
    acyl glycinates, 59
    acyl sarcosinates, 58
    acyl taurates, 57–58
    hydrolysed proteins surfactants, 60
    other amino acid surfactants, 59
sweat baths, 104
synthetic surface active agents, 29–100
    amphoteric surfactants, 31–35
    anionic surfactants, 35–60
    cationic surfactants, 67–70
    nonionic surfactants, 71–91

T

taurates, 59, *see also* acyl taurates
taurine, 27, 45
    alkyl, 31
    N-methyl, 57, 58
taurocholic acid, 27
Theodorus Priscianus, 103
Th. Goldschmidt, 31–32, 97

V

Versailles 13

W

washing and cleaning habits, 103–116
    body and personal hygiene, 105–116
    linen and clothes, 103–105
Wurtz synthesis, 42, 79, 89, 91

Z

zinc, 42
    halides of 60
zinc-copper alloys, 84

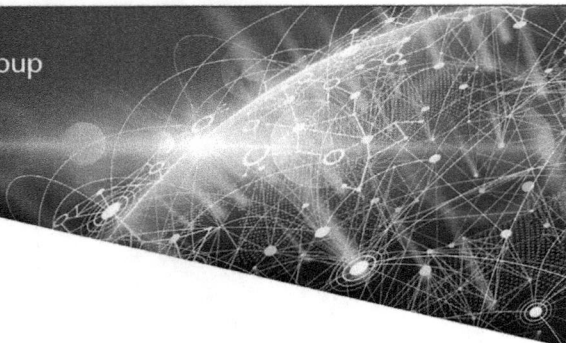

For Product Safety Concerns and Information please contact our EU
representative GPSR@taylorandfrancis.com
Taylor & Francis Verlag GmbH, Kaufingerstraße 24, 80331 München, Germany

www.ingramcontent.com/pod-product-compliance
Lightning Source LLC
Chambersburg PA
CBHW060319220326
41598CB00027B/4369